AF843205

Quantum Finance

With its groundbreaking approach to financial management under dynamic conditions, this book equips readers with a new perspective on financial systems, moving beyond classical models to embrace the inherent uncertainties of markets.

Since they assume a static environment, many traditional financial models fall short of capturing market dynamics, leading to poor outcomes. This book draws parallels between quantum mechanics and financial management to explain the erratic behavior of the world of finance. It uses quantum concepts including uncertainty, superposition, and entanglement, which have been nurtured over time by renowned physicists. But rather than a deep dive into quantum mechanics or computations, it presents simple principles and recent case studies to illustrate them. The book's fresh analogies and illuminating thought experiments explain complex financial phenomena and guide readers toward the flexible, dynamic mindset required for modern-day financial decision-making.

This book is a must-read for finance professionals, corporate leaders, investors, and policymakers looking to navigate the unpredictability of the modern financial world.

Arif Ahmed is a chartered accountant in the UK (ICAEW) and India (ICAI). He is also a management accountant from the UK (CIMA) and a chartered global management accountant (AICPA-CIMA, USA). He holds an MBA and a doctorate in finance. He has over 35 years of post-qualification experience of working with clients from diverse industries and regions.

Quantum Finance

The CFO's Survival Kit for an Uncertain World

Arif Ahmed

Routledge
Taylor & Francis Group
NEW YORK AND LONDON

Designed cover image: Getty

First published 2026
by Routledge
605 Third Avenue, New York, NY 10158

and by Routledge
4 Park Square, Milton Park, Abingdon, Oxon, OX14 4RN

Routledge is an imprint of the Taylor & Francis Group, an informa business

© 2026 Arif Ahmed

The right of Arif Ahmed to be identified as author of this work has been asserted in accordance with sections 77 and 78 of the Copyright, Designs and Patents Act 1988.

All rights reserved. No part of this book may be reprinted or reproduced or utilised in any form or by any electronic, mechanical, or other means, now known or hereafter invented, including photocopying and recording, or in any information storage or retrieval system, without permission in writing from the publishers.

Trademark notice: Product or corporate names may be trademarks or registered trademarks, and are used only for identification and explanation without intent to infringe.

ISBN: 978-1-041-02514-6 (hbk)
ISBN: 978-1-041-01977-0 (pbk)
ISBN: 978-1-003-61963-5 (ebk)

DOI: 10.4324/9781003619635

Typeset in Sabon
by Apex CoVantage, LLC

We are all entangled, so you all are here
Some we can see, some we cannot.

Contents

Preface

> Ah, but a man's reach should exceed his grasp, or what's a heaven for?
>
> – Robert Browning

Human knowledge seeks to expand its reach to remain relevant. Finance is no exception to it. Over the years, finance has witnessed many changes – some cosmetic, some fundamental. Over the years, one of the common allegations against finance involves its failure to foresee major failures or explain unprecedented growth. These have been classically treated as rare events – as outliers. In fact, they are considered so rare that we ignore them.

This is where the book comes in. The laws of quantum physics treat similar phenomena as rare but an inherent possibility within system behavior – recognizing it as a matter of probability and not impossibility. Quantum physics has influenced computational finance and investment management. But they did not explore whether the business resembles the quantum world, and that is why the existing financial concepts that express the future as a single predicted value fail to explain many sudden developments. The fact is that the stability that we see is only one of the possible options. This book attempts to encourage finance professionals to accept uncertainty as a fundamental property of the financial world – and not treat it as a problem that needs to be solved.

It will be wrong if I say that business professionals today do not practice what the quantum thought process would have recommended. However, this alignment is mostly accidental – it has happened without consciously recognizing the underlying quantum nature of financial systems. The examples point out that those actions were great quantum practices, deliberate or not.

At the same time, this book is not a textbook that explains all concepts, nor does it have prescriptive text. This is a book of possibilities that you can explore while deciding on a strategy. It represents finance's attempt to reach beyond its current grasp – exploring quantum possibilities that classical models cannot yet fully comprehend.

The book doesn't claim to invent quantum finance. But I will stick my neck out to say that this is a pioneering effort in finding quantum behavior in so many aspects of finance, related business practices, and regulations.

Of course, I will claim to have found Quantaland.

Acknowledgments

This book draws on the efforts put in by so many people over many years to develop the central theme. I should mention Dr. Jane Alam not only for his efforts in explaining the concepts of quantum principles but also for inspiring me to consider how the world of finance may resemble the quantum world in some way. I thank Megha Nagpal from Informa, Dubai, for reaching out to the publishers – that small little step that makes the journey possible. Thank you, Meredith Norwich for believing in the concept and Bethany Nelson for guiding us all through. I hope you never counted the number of times I requested an extension of the deadline for submitting the manuscript. I must acknowledge the use of various tools and resources to research the topic and gather facts for use in case studies. I must thank Veena Hingarh and Arnaaz Ahmed for taking the load off me from other books and allowing me to complete the manuscript – besides providing useful tips. I am thankful to Partha Choudhuri, a dear friend and a career regulator, for acting as the sounding board for some of the ideas.

I am sure I am missing a few who should be acknowledged. Thank you all.

How to Use This Book

Everyone has limited time, so we prioritize accordingly. This is true for the readers, too. So I have split the book into sections focusing on various aspects of business and financial decision-making. If you're in a hurry, read the first two chapters and then the rest in your order of interest. You will not find any difficulty in following them. To achieve that, I had to explain similar phenomena across various business domains in separate sections. Once you complete reading the book, you will recognize how the fundamental properties of quantum finance are manifested in various behaviors of the world of finance. After all, requiring a fixed reading sequence would contradict the very quantum principle this book advocates: multiple pathways can coexist simultaneously until your specific needs collapse them into the optimal reading pattern for your situation.

Quantaland examples demonstrate quantum finance principles through fictional companies facing real business challenges. These scenarios explore various strategic options without prescribing "correct" answers. The reason is that in quantum finance, optimal solutions depend on specific circumstances and timing, which you can only measure in your situation.

The case studies under "Meanwhile, far from Quantaland" are all incidents from the real world. Resemblances of these decisions to quantum thinking are my personal interpretations; I am not claiming nor disputing that the entities cited in these examples actually had such a thought process. The entities involved have not been approached to validate the interpretations.

The book focuses on the application of quantum theories to various aspects of finance and associated functions. Each chapter focuses on a specific function and discusses how the quantum principles apply there. That is why you will find short, concise paragraphs explaining the central idea, leaving the discovery of precise tools up to you. This has led to repeated discussions about the same quantum principles, but each discussion has different end objectives.

The examples are not recommended solutions to a problem. Their role is to support your thought process. This approach honors your professional expertise by providing conceptual frameworks and quantum-inspired thinking tools, then trusting you to adapt these insights to your unique circumstances. After all, you understand your organization's specific quantum state better than what any external prescriptions could capture.

Somewhere beyond the horizon, Quantaland does stand
Where funny things unfold, in a whimsical and strange land.
Perhaps it is meant to be so, though we view it differently
A world where odd's the norm, and wonders roam free.

Elephants soar with wings over forests so blue,
Where numbers follow no order, giving us no clue.
Chickens lay eggs with care and climb inside to be born
Under Quantaland's wild ways, our logic comes undone.

The sun rises in the North and sets with the wind's flow,
Pearls from seas flow in streams, follow its gentle glow.
Apples sprout on orange boughs, sunflowers change their hue,
In topsy-turvy Quantaland, what's strange is always true.

Shoes dance on their own, the king is picked by the crown.
Students slip through blackboards to wear their teachers' gown.
A mad, mad world it is, where logic is held at bay,
In Quantaland's utter chaos, that's where we live and play.

Chapter 1

Introduction

"*The fishing cat knows that flowing water holds more fish than frozen streams*," goes an old Quantaland proverb.

Picture this: You're the CFO of a profitable manufacturing company. Your quarterly numbers are solid, and the board is pleased. Then, overnight, a supply-chain disruption in Southeast Asia, a shift in consumer preferences, and a regulatory change push your stock price down by 23%. The financial models that predicted steady growth for the next three years suddenly seem worthless.

Sound familiar? If you've worked in finance for more than a few years, you've lived through some version of this story. The 2008 financial crisis, the COVID-19 pandemic, and the supply-chain chaos of 2021 are all witnesses to how our carefully constructed financial models failed to capture the interconnected, dynamic nature of modern markets.

The issue isn't that we lack technical proficiency. Finance professionals are brilliant at building sophisticated models, conducting thorough due diligence, and calculating precise risk metrics. The problem is we're trying to navigate today's dynamic, hyper-connected markets with yesterday's static thinking. We're trying to predict a complex, interconnected system by taking snapshots of individual components.

What if there was a better way? What if, instead of treating uncertainty as something to eliminate through better forecasting, we embraced it as fundamental to how markets actually work? Could we construct financial strategies that thrive on volatility instead of succumbing to it?

Mastering this requires three fundamental shifts in how you approach finance. First, you'll discover why treating financial systems as dynamic, interconnected networks yields better insights than static analysis. Second, you'll explore how quantum concepts such as superposition and entanglement provide powerful frameworks for understanding market uncertainty and interdependence. Last but not least, you will discover how these quantum-inspired methods can change the way you approach risk management, strategic planning, and investment analysis. By the

DOI: 10.4324/9781003619635-1

end of this chapter, you'll start seeing financial challenges through a different lens.

The essence of quantum finance is to replace the futile quest for perfect prediction and build strategies that are powerful because they are adaptable.

Finance is quantum

Our first step should be to redefine the financial and its reporting system as dynamic processes – where everything interacts – rather than as static snapshots. Over the years, the finance world has believed in using static values, though the underlying process has always been dynamic. In the world of traditional finance, we compute market volatility and then use it to make a decision. One could argue that a static value is necessary for making a decision. If I am computing the break-even point for a project, I need to freeze certain values to arrive at the break-even point. Yes, but we don't think it's the only way to measure the break-even point. We can view the process as a continuous journey, where both the break-even point and our performance undergo constant evolution. So should our metrics. Historically, dynamic financial behaviors were estimated using static metrics – something like using a still image to analyze the speed of a racing car. Today, with advances in data processing and analytics, we can better capture these dynamic interactions in real time and use them in our analysis and planning.

So, what do we do with finance? Simply put, we introduce a certain amount of capital into a potential revenue-generating activity. It flows into the activity, and we hope that it grows while interacting with other elements in the activity chain. We need to introduce a minimum amount of finance for it to be able to grow by interacting with other elements. If it grows, at the end of the activity we will have more finance, or resources, than we started with. Incidentally, we do not need to wait till the end of the activity; we can withdraw the finance at an earlier time, though this depends on the terms of participation.

To understand this better, consider an analogy from an unlikely place. Have a look at what happened in Los Alamos, New Mexico, between 1942 and 1945. A group of scientists was trying to introduce neutrons into fissile material like uranium-235. These neutrons interact with atomic nuclei, causing them to split and release energy along with more neutrons. This created a chain reaction where the output (energy and neutrons) exceeded the initial input. To sustain a chain reaction, a discrete, minimum number of neutrons is required. This is called "critical mass." Below this threshold, the reaction fizzles out; above it, each fission event produces enough neutrons to trigger more fissions, creating a self-sustaining chain reaction.

Business ventures operate on the same principle. They need a critical mass of capital to ignite a chain reaction of returns. However, unlike nuclear fission, where the laws of physics are well defined, financial "chain reactions" depend on market conditions, timing, and countless human decisions, which make them extremely complex and unpredictable.

Nuclear fission itself is a quantum mechanical process where energy is released in discrete packets (quanta) when atomic nuclei split. The energy levels in atoms are quantized, and the fission process involves quantum tunneling, which is the ability to pass through energy barriers and probability distributions. Quantum tunneling introduces an element of probability. There is always a chance, however small, that a quantum event will occur even when classical physics says it's impossible.

The similarities here are striking, aren't they? To avoid confusion, we should clarify that the words "quantum mechanics" and "quantum physics" are used interchangeably. Keep looking into it, and you may find the big bang theory of creating wealth. Just like nuclear physics requires triggering a chain of energy-releasing reactions, finance involves triggering chains of value-generating reactions. We will explore this parallel further.

Overview of quantum finance and why finance resembles quantum systems

As we established, wealth creation starts when capital is introduced to a potentially profitable activity. Financial statements track the progression of the process. Let us look at the journey that can be titled "*from cash to cash*." Wealth is generated only when the end cash is greater than the cash at initiation. Of course, cash is a generic description of any asset that is extremely liquid with stable value and unencumbered. Next, let us use some financial figures and investigate the process deeper.

As the cash is introduced to the economic activity, it hits the balance sheet, which recognizes the cash as an asset and the corresponding owners' equity. Now the cash is utilized to purchase assets. These are commonly of two types. One type helps in production, and the other type is consumed in production. The former is expected to be used over a longer period, and the latter is consumed within an operating cycle. This is reflected on the asset or application side of the balance sheet. A part of the initial cash is converted into these assets, fixed and current, as we call them. If a non-owner supplies finance, we recognize this financing on the source side of the balance sheet as a liability, which is distinct from the owners' equity. The fund provided by them receives the same treatment on the asset or application side of the balance sheet.

Next comes the second key player: the income statement. The assets flow from the balance sheet into this statement. The cash is used to fund various expenses unless they are supplied on credit. The current assets are mostly consumed; a small part may remain to be used in the next cycle. The fixed assets are used up gradually over their expected life, and a notional cost is charged to the income statement as depreciation. If the fixed asset is intangible, the same process is called amortization. We will discuss the implications of depreciation in a while. Cash comes when what we produce is sold, either immediately or after the stipulated credit period. If income exceeds expenditure, we have created wealth; if not, we have destroyed it. This wealth is then transferred back to the balance sheet, where it is either distributed between the suppliers of capital or retained for later use. The income statement is reset for the next cycle.

The end-of-activity balance sheet shows cash and other assets. We expect to realize at least the reported values from the assets. If the business is to be terminated, all assets are disposed of, liabilities are settled, and we are left with a balance sheet with cash and capital. The distribution from the profit is the input to the creation of wealth in the hands of the owner.

I promised a discussion on depreciation. This is a mechanism to convert the fixed assets back to cash, the income statement being the facilitator. We have used cash to procure the assets, and we need it back. Depreciation shows the way. We know that depreciation is a notional expense; it does not involve any cash outflow. By charging depreciation as an expense, we have reduced profit. Thus, the owners' right to receive a return is determined by post-depreciation profit under the accrual system of accounting. If we consider an all-cash operation, the cash surplus will exceed the profit by the amount of depreciation. The entire cash surplus will be there as an asset. Even if we distribute the entire reported profit as dividends, the remaining cash will equal the depreciation value. So what has happened? We have reduced the value of the fixed assets by the amount of depreciation and have increased cash by the same value. Assets are being diluted into cash.

Consider this example from Quantaland, the faraway land, where they use the Q$. In this example, we have fixed assets worth Q$500. Profit after tax is Q$200, while the cash generation during the period is Q$250. The difference of Q$50 is caused by the depreciation; it has been treated as an expense under the accrual system. The closing cash balance has gone up by Q$250, while the accumulated profit has gone up by Q$200. So where does the balance of Q$50 come from? Look at the fixed asset, which has reduced from Q$500 to Q$450, a dilution of Q$50. The Q$50 investment in fixed assets has been recovered through operational activities. Depreciation does not generate cash flow directly, but it allows cash generated from operations to be retained rather than being paid out as profit. Over time, the fixed asset is "used up," and its book value declines due to depreciation.

Table 1.1 Depreciation and cash flow

Statement of income and expenditure		
	Accrual	*Cash*
Sale of products	1000	1000
Cost of goods sold	600	600
Finance cost	100	100
Depreciation	50	—
Profit/cash before tax	250	300
Tax	50	50
Profit/cash after tax	200	250
Statement of sources and application		
Application	Initial	Close
Fixed asset less depreciation	500	450
Cash	500	750
Total application	**1000**	**1200**
Sources		
Capital	500	500
Accumulated profit	—	200
Borrowings	500	500
Total sources	**1000**	**1200**

The cash invested in the assets is recovered gradually. In effect, depreciation acts as a conduit for transforming a fixed asset into free cash.

Therefore, to create wealth in business, we set into motion a chain of reactions arising out of the interaction of various assets. This wealth-creation activity may be disrupted, accelerated, or enhanced by withdrawing fixed assets, consuming the asset at a faster rate, or bringing in new assets. The wealth that is created comes in discrete intervals. The withdrawal, acceleration, or introduction can also be done in a discrete interval. One may argue that financial transactions can happen in any amount. While this may be theoretically correct, in practice, businesses operate at defined levels of investment. This could be defined by production capacity levels, packing sizes, and other related factors that result in discrete values. The possibilities are clearly defined, but they may not follow a sequential order.

As our discussion shows, finance operates a lot like a quantum system. This quantum nature of finance creates a fascinating bridge between quantum mechanics and financial theory. This opens up an entirely new way to approach financial decision-making. This approach uses quantum principles to model financial decisions in ways classical finance often overlooks. Here's a quick rundown of the main concepts in plain English.

Quantum mechanics in a nutshell (or how to pretend to know about it)

Quantum mechanics is the study of matter and energy at the smallest scales such as atoms and particles. However, the focus shifts from the particle itself to its wave-like path. Once we think dynamically, we can no longer pinpoint a single position. The particle could be anywhere on that wave. Thus, we enter the realm of probability, leaving static definitions behind. This is where it gets intriguing. Let's now break down the major concepts we'll be using throughout the book.

- **Superposition**: We tend to think that an element is at some place, instead of thinking of a particle having a definite path or location. The quantum theory describes it as being in a superposition. This is a blend of a range of possible positions or states until it is measured. This leads to the idea that matter and energy can exist in multiple states simultaneously. This should not be understood to mean that the particle exists in multiple locations at the same time. Rather, its position isn't well defined until we observe it. Until then, it is described by a probability distribution that spans multiple positions. The state we observe depends on the act of measurement. We will observe only one outcome, despite the possibility of many others. A rough analogy is a spinning coin. While it's in the air, you can't say for sure whether it is heads or tails. While it's spinning, we can't predict whether it will be heads or tails; it could be either. However, once it lands and you observe it, the outcome is determined as either heads or tails. Similarly, a quantum particle exists in all states simultaneously until measured. This uncertainty is fundamental to quantum mechanics and not because of a lack of information. That's the heart of superposition. This bizarre idea is famously captured in Schrödinger's cat thought experiment – a concept we'll dive into later.
- **Entanglement**: This refers to a situation where two particles become so deeply connected that the state of one instantly determines the state of the other. The distance between them is of no consequence. When particles are entangled, measuring one immediately tells us something about the other. Crucially, there's no classical communication, like a signal of any type, between them. Instead, their properties are linked in such a way that when one is observed, the other's state is known instantly. While it's challenging to construct a perfect everyday analogy, imagine two shoes wrapped separately: if you open one wrapper and identify the right shoe, you intuitively know the other wrapper contains a left shoe. However, quantum mechanics goes one step beyond this. The shoes lack a distinct left or right until you open a single box. They acquire a definite identity only when one wrapper is opened. However, it is important to note that entangled particles do not literally possess a definite state until

they are observed. Entangled particles do not necessarily have the same value. In fact, their values are interdependent in a well-defined way. This interdependence is not just a statistical correlation, like predicting the sale of butter based on the number of loaves of bread sold. This interdependence is a deep, physical connection that defies classical explanation. That's what makes entanglement such a unique and puzzling feature of quantum physics. If this seems confusing, please don't be concerned. It baffled Einstein, too. He even called entanglement "spooky action at a distance."

- **Uncertainty Principle**: Known as the Heisenberg Uncertainty Principle, this states that one can never simultaneously know both the position and the momentum of a particle with absolute precision. The more precisely we determine one, the less precisely we can know the other. This goes beyond mere limitations of our tools or how we look. It is a fundamental property of quantum systems. Think of it like an F1 car on a track. If you try to calculate its speed, you need to measure how far it travels over time. This gives you a good idea of momentum but not its precise position at a single instant. If, instead, you try to capture its exact location at a specific moment, you lose the ability to calculate its speed precisely. Consider trying to understand both the exact current price of a volatile stock and its exact rate of change at the same instant. The faster the price moves, the harder it becomes to pinpoint where it "is" versus where it's "going." These are just rough analogies from our classical world, but they capture the essence of the trade-off. In quantum mechanics, however, this uncertainty is not due to observational difficulty. It is built into the very nature of particles. This is why Heisenberg's principle is so revolutionary.
- **Wave-Particle Duality**: An important reason quantum theory is so puzzling is because particles like photons and electrons defy our everyday categories. They can behave as waves in some situations and as particles in others. This dual behavior isn't about revealing hidden traits. It's that the very act of measurement defines what we see. When we set up an experiment to detect wave-like properties, these entities spread out, interfere with each other, and form patterns like ripples on water. But when we design an experiment to detect particles, the same entities appear as discrete, localized impacts with individual hits at specific points. Remarkably, it's the same underlying entity each time; the experimental setup determines the outcome. Think about watching a movie being filmed versus watching the final edited version. During filming, you see continuous action with actors moving, cameras rolling, and the whole scene flowing like a wave of activity. But in the final edit, that same footage appears as discrete cuts, specific frames, and individual scenes like particles of storytelling. The key insight? The editing process, similar to a quantum measurement, not only reveals reality but also

creates it. It actively transforms the same source material into a different reality. The raw footage and the final cut are two different realities built from the same events. In this sense, the measurement method, in addition to offering different views, actively determines the behavior we observe. This involves more than switching between two preexisting states. Rather, the act of observation plays a vital role in shaping the reality we encounter. This principle is crucial when we apply quantum thinking to markets. The tools we use to analyze data can actually change the behavior we're trying to measure.

- **Quantum measurement**: This is the process of observing or measuring the state of a quantum system. It represents one of the most fundamental and puzzling aspects of quantum mechanics. In classical physics, measurement simply reveals what's already there. Quantum measurement, by contrast, actively disturbs the system it observes. Before measurement, a quantum system exists in superposition, which is a blend of all states, each associated with a probability. The moment we perform a measurement, this superposition collapses into one definite state. The system transitions from a set of potential outcomes to a single, observable value. In most practical cases, the system cannot return to its original superposed state. The measurement has permanently changed its quantum nature. Here's the crucial part: the result of any single measurement is fundamentally probabilistic. We can use quantum mechanics to predict odds, not certainties. But it can never tell us exactly which outcome will happen. However, once measured, the result is final and definite. Take a complex derivative. Its value hinges on multiple market factors at once. Before you "measure" its value by referencing live market prices, the derivative exists in a notional state. At this stage, all potential valuations coexist, each tied to a different underlying scenario. Price discovery – whether through trading, marking to market, or using valuation models – collapses that range of possibilities into one realized outcome. Different valuation methods can provide different, yet equally valid, answers. Therefore, how you choose to measure not only reveals the value, but it can also actually help determine it.

These concepts posed a challenge to classical physics and paved the way for innovations such as semiconductors, LEDs, and MRI machines. We will now examine how these concepts are relevant to the world of finance, not only to investments but to the entire financial ecosystem.

Financial markets as quantum systems

So far, we've defined a quantum system as one governed by the bizarre rules of quantum mechanics. It describes how matter and energy behave at the atomic and subatomic level. But it's not just about size; it's fundamentally about uncertainty, superposition, and probability. Classical systems, in contrast, are considered deterministic, following a smooth, continuous path. Quantum systems are probabilistic at their core – their outcomes can't be predicted with certainty. Even when classical systems incorporate probability, like in stochastic models, they differ fundamentally from quantum systems. In quantum mechanics, probability isn't about randomness; it emerges from deep principles like superposition and entanglement. This is what sets quantum systems apart from the probabilistic models we use in finance.

After discussing the fundamental quantum ideas, we will now examine if financial systems truly exhibit these peculiar quantum behaviors. Do financial states exist in superposition? Is there a financial equivalent to quantum entanglement? Does price discovery reduce possibilities to a single reality like a wave function? Is market unpredictability fundamental like quantum uncertainty? We'll also explore how small, overlooked micro-events can trigger massive macro outcomes. Our goal is to search for all outcomes, not just the single, deterministic number a financial report provides us. Furthermore, we will examine previous instances where traditional finance failed due to its disregard for these intricate interactions.

The financial system primarily does three things:

- First, it provides a framework to measure value and track wealth creation.
- Second, it manages the mechanisms for exchanging value between participants.
- Finally, it maintains a system to record transactions and report who holds what wealth.

Take a running shoe factory, for example. Here raw materials are converted to finished shoes using people, machinery, logistics, and utilities. Countless activities ranging from design to demand forecasting add value during this conversion process. Marketing, sales, and delivery create a system that facilitates the exchange of value between participants. This system transfers the values between two participants in the financial system. The seller gives away the shoes and the buyer pays for them from an existing pool of wealth. Finally, accounting records all of this to document who owns the new wealth and tracks its transfer.

Now consider a financial product: an option to buy Q$1 million in 30 days. The option writer creates a call option and receives a premium payment. The option buyer gets the right to exercise the option on the decided date or within a time limit. Upon exercising the option, the buyer can either get delivery of Q$1 million by paying the agreed price or settle it on a net basis by receiving the difference between the contracted price of Q$ and the market price of Q$ on the transaction date. This is where it gets intriguing! The buyer calculated her risk and decided the premium was a fair price to mitigate it. The financial market provided the design of the instrument and the settlement process. The accounting system recorded the wealth transfer and the removal of the risk. The quantum-like features here are difficult to miss. The option embodies multiple future states at once: it could be profitable, worthless, or anything in between. This superposition collapses into a single reality the moment a decision is made.

A massive number of activities make both of these examples possible. Unfortunately, most of us only see the financial system through the narrow window of financial reports. They show the picture after the activity has ended. Or, more precisely, they report on a sub-activity that has been completed. However, we use these reports about the past to make decisions about the future.

Accounting is crucial for the financial system to function. But we must recognize its two defining features.

- First, it links wealth and its creation to ownership and control. If a company benefits from wealth not owned or controlled by it, those won't appear on its financial statements. So, if a store benefits from foot traffic generated by a neighboring business, that benefit isn't recorded anywhere. Report users can completely miss what's actually driving success.
- Financial reports bind activity to a specific period, the accounting year. We record wealth creation in the period it occurs. A firm decision to launch a major project next year won't appear in this year's reports. The market might price this in, but accounting rules omit it.

These two features force us to view everything through the lens of ownership, control, and a fixed period. We instinctively ignore anything that does not belong to us, is not under our control, or falls outside the current period. This means we try to understand a dynamic, interconnected system with static summaries, which severely limits our perspective.

Furthermore, it's critical to understand that more information can't eliminate the inherent uncertainty in wealth creation. This uncertainty is not caused by a lack of data; it is a fundamental feature of the system. It mirrors the fundamental uncertainty we see in quantum systems. No amount of forward-looking information in a report can make this uncertainty disappear. Better information might help us map the odds more accurately, but the underlying uncertainty remains.

To understand value creation, you have to look beyond the financial reports. Financial reports only show us a snapshot of a link in the value chain. Multiple sub-activities contribute to the completion of the activity. These activities have their own paths and may depend on some sub-activities elsewhere. Our ability to collect payment depends on our customer's ability to pay, which depends on their customers paying them. If you stop thinking in terms of single companies and start thinking in terms of chains, you see a vast network spread across the entire economy. Quantum finance helps you understand how a financial system relates to these vast, interconnected chains of activity.

Quantum theory might seem counterintuitive, but its mathematical structure offers a powerful way to model the complexities of finance. While classical finance presumes determinism and rationality, real markets are characterized by uncertainty, intricate interconnections, and the dynamic nature of observation. We will now map five key quantum concepts directly into finance. As you would have noticed, we've already introduced these concepts.

- **Superposition**: Superposition means a system that exists in multiple states simultaneously until measurement forces it into one specific state. In finance, this resembles how investment opportunities exist across multiple potential outcomes, such as profitability, breakeven, or loss-making, until market forces "measure" them through actual trading or business results. A stock option embodies this perfectly: it simultaneously holds the potential for significant profit, modest gain, or total loss until exercised or expired.

  Imagine a quantum algorithm that, along with assessing portfolio allocations one by one, also evaluates them simultaneously as probability amplitudes. Unlike a classical model that searches for a predetermined "best" option, the quantum system genuinely exists in a superposition of all possibilities. This allows it to identify optimal risk-return profiles that classical computers could miss because it can process an exponential number of combinations in parallel before the final measurement collapses it into a single, optimal solution.

  Have a look at the flexible assembly line of Quantaland Auto's, which can produce three different models. A classical approach would analyze the cost and efficiency of each model in isolation. A quantum-inspired view, however, sees the line's true potential in the superposition of all production states. This perspective uses the mathematical principles of quantum mechanics to identify hidden efficiencies in the interactions between models. These are optimizations a sequential analysis would never see. This superposition persists until the production order arrives. That order acts as a "measurement," collapsing all those simultaneous possibilities into the single reality of that day's production schedule. The flexibility of the system itself creates this quantum-like state of potential.
- **Entanglement**: Financial markets are a web of complex interdependencies much like the "spooky action at a distance" seen in quantum

entanglement. Think about when a central bank changes rates. The effect isn't a slow ripple; it's an instantaneous repricing of assets globally, a correlation that feels quantum in its speed and scope. It's an instantaneous repricing of assets globally, driven by shared psychology and automated systems, a correlation that feels quantum in its speed and scope. While it's not literally quantum entanglement, the mathematical frameworks used to describe entanglement are surprisingly effective at modeling these market interconnections. Consider a just-in-time manufacturing ecosystem in Quantaland. Company A makes chips, Company B uses them for sensors, and Company C builds cars with those sensors. Their zero-inventory contracts create tight, mathematical dependencies that entangle their operations financially. Contractual logic – not just communication – instantly links Company A's production state to Company C's assembly targets. If a quality issue halts Company C's line, the "measurement" of this failure is not a simple transmission of a message. It has a greater impact and forces an instantaneous, mathematical adjustment to production parameters at Company A's facility, regardless of distance or direct contact. The entire network's range of production states collapses into a single, new reality based on that one event. This kind of deep, instantaneous interdependency is precisely what quantum-inspired math is designed to model. In practice, smart contracts execute the adjustment, but the underlying dependency structure behaves like entanglement. Changes in one node force changes in others due to the system's design, not as a secondary effect.

- **Uncertainty Principle**: Financial analysis is riddled with trade-offs that feel eerily similar to quantum uncertainty. The more laser-focused you are on minute-by-minute price ticks, the blurrier the long-term trend becomes. This isn't a fundamental law of physics like Heisenberg's principle but a practical reality born from market complexity and our own analytical limitations. Yet, using quantum-inspired math to model, this trade-off can give us a powerful new lens. High-frequency traders, for example, excel at capturing micro-movements but often completely miss the macro picture. This is the financial analog to quantum uncertainty, where maximizing precision on one variable (price) inherently reduces it for another (time horizon).

Let us look at another example. Take Quantaland Engineers, a precision machinist. Their quality control is incredibly precise, measuring components to within microns. But this rigorous inspection slows down the production line. Here's the trade-off: the more they try to eliminate uncertainty about quality, the more uncertainty they introduce into their production timing. Optimizing for one forces a compromise on the other. This isn't just a scheduling problem; it's a fundamental feature of their operating system.

- **Wave-Particle Duality**: Quantum entities display wave-like behavior when unobserved and particle-like traits when measured. In markets, we see this duality in how we analyze behavior. We look at overall market sentiment, a wave-like phenomenon, and individual trades, which are discrete, particle-like events. But the act of widespread observation, like a prominent analyst report, can itself influence that collective sentiment, effectively "collapsing" a trend of possibilities into a specific market outcome. This same duality applies to manufacturing data.

  At Quantaland Engineers, data itself have wave-particle duality. Engineers can view production through a wave lens using tools like throughput analysis to see smooth patterns, rhythms, and efficiencies that ripple through the system. From this perspective, production lines interfere with each other like waves, creating bottlenecks or synergies. Alternatively, they can view the same system through a particle lens using discrete event simulation to analyze individual work orders, machine failures, or quality rejects as distinct, impactful events. A rush order (a particle) can disrupt a smooth production flow (a wave). The crucial insight is that the choice of tool goes beyond revealing different views to actually determining which behavior becomes relevant for decision-making. The data themself behave as either a wave or a particle based on how we choose to measure it.
- **Quantum Measurement**: The best financial analogy for quantum measurement might be in algorithmic trading. A classical trading algorithm analyzes the market and then selects a preprogrammed strategy such as buying stocks, selling bonds, or hedging currency. A quantum-inspired system works differently. It maintains multiple strategies in a simultaneous, active state of readiness, demonstrating true superposition. All strategies are "live" and ready to be executed. The system avoids choosing between them until the very last moment. The arrival of market data acts as a "measurement," causing this superposition of active strategies to collapse instantaneously into a single, executed action. The decision is created at the moment of measurement. Let's understand this in greater detail.

  Here's the key difference. A classical algorithm calculates probabilities for each strategy and then picks the winner. It's a selection process. A quantum-inspired algorithm holds all strategies in a state of potential where they can interact and interfere with each other until the moment of execution. The market data transcend simple selection by interacting with this entire field of available strategies, potentially giving rise to a hybrid or novel decision that wasn't preprogrammed. It's not choosing; it's creating a response in real time.

  Thus, in a classical system, when the interest-rate announcement is made, the algorithm calculates, "Given this data, the strategy of buying stocks has a 70% success probability, the strategy of selling bonds has 45%, and the strategy of hedging currency risk has 30%." The system selects Strategy A as the "best" choice and executes the predetermined optimal strategy. The decision was not discovered at that point in time; mathematically, it had always been there. The market data revealed which pre-calculated option was optimal.

In a quantum system, before the market announcement, the system maintains a superposition of the three strategies, each with its probability amplitude of α (0.7), β (−0.5), and γ (0.3). The market announcement acts as a quantum measurement, triggering the decision-creation process. The market information moves beyond merely selecting from A, B, or C. Instead, it interacts with the superposed state (α, β, γ) to create a new decision that incorporates elements from multiple strategies. The system might execute a hybrid action like "Buy stocks while simultaneously establishing currency hedges." This decision wasn't pre-calculated but emerged from the quantum interaction.

The quantum-inspired system processes information in fundamentally different ways, not merely faster or more efficiently. It processes information in a fundamentally different way, where the act of receiving market data becomes part of creating the trading response.

This concept extends to manufacturing. Imagine a circuit board on the line at Quantaland Electronics. Its final quality isn't a fixed, hidden property. Instead, it exists in a superposition of potential states such as "premium," "standard," and "rework" until the moment of inspection. Measurement techniques – like the calibration of the tester and the ambient temperature – actively participate in determining quality rather than passively revealing it. It actively participates in determining the final outcome. The same board might pass in the morning but be marked for rework in the afternoon based on subtle changes in the measurement environment. This means quality control isn't only about making a consistent product; it's about understanding and controlling the entire measurement ecosystem.

Of course, integrating quantum principles into finance isn't simple. Most applications so far have been in forecasting, not in reinterpreting everyday business activities. A true understanding requires a full rethink of our traditional models for wealth creation, value transfer, and reporting. A key objective of this book is to shed light on these alternative interpretations.

Where classical finance breaks down

Classical financial models have been the workhorses of the industry for decades. However, the intricate and interconnected nature of today's business world has challenged these models, revealing significant limitations. These limitations include the following:

- **Simplistic assumptions**: Classical models frequently rest on certain assumptions. They include rational investor behavior, normal distribution of returns, static cost structures, and so on. In the real world, investor psychology, market shocks, and dynamic costs introduce significant variability, making the static assumptions redundant.
- **Linearity bias:** The linearity assumption and rule-based behavior are among the basic foundations of many traditional models. This structure does not recognize the nonlinear dynamics observed in financial markets, especially during periods of volatility or systemic stress. While a limited number of recent models incorporate nonlinearity, they primarily focus on investment decisions.
- **Inadequate treatment of uncertainty**: Classical models often quantify risk without any awareness of unknown events or factors. For instance, you may prefer a project with a lower break-even point without considering the probability of reaching it.
- **Model scalability and data volume**: The acceptable performance of some models when using small datasets often hides their inability to manage complex data. For example, the number of asset pairings in portfolio models increases combinatorially, making manual or linear analysis impractical beyond a certain threshold.
- **Static decision frameworks**: Classical decision models often assume fixed objectives like profit maximization. It ignores the fact that strategic goals may evolve. For instance, pricing models used in early-stage market penetration may differ significantly from those applied in mature markets. Using the same model to evaluate a disruptive startup and a stable blue chip is a fundamental category error.

To be fair, classical models have evolved, incorporating tools such as Monte Carlo simulations and behavioral insights. However, these sophisticated tools encounter challenges when confronted with genuinely intricate, interconnected systems. They struggle with the sheer computational complexity of multi-asset, multi-factor environments.

This is where quantum-inspired approaches offer a potential way forward. The goal isn't to replace financial theory but to solve these specific computational and modeling bottlenecks. They also better account for the "observer effect," where a large trade itself moves the market it's trying to predict. They do this by baking these feedback loops directly into the model, not treating them as external anomalies. Their potential power lies in processing thousands of interdependent variables at once. This makes them suited for problems such as environmental, social, and governance

(ESG) portfolio optimization, high-frequency trading, or assessing systemic risk.

Schrödinger's cat and financial decisions

It's now time to use a famous quantum thought experiment to illustrate this shift in thinking. Schrödinger's cat serves as a perfect illustration of superposition and how observation defines reality. In the experiment, a cat is in a box with a radioactive atom that has an equal chance of decaying and releasing poison. Until the box is opened, the state of the cat remains unknown – whether it is alive or deceased. Therefore, until observed, the cat's state is genuinely undefined. It is not that it's simultaneously alive and dead but that the system exists in an unmeasured state where the outcome hasn't been determined. This superposition collapses into a definite reality only when observed.

This is a powerful metaphor for finance. It shows how quantum-inspired thinking oversees uncertainty differently. The point isn't that your stock portfolio is both up and down. It's that the analysis can hold all those potential scenarios as simultaneously real until a decision is forced.

This analogy forces us to recognize multiple potential outcomes at once and adopt a more probabilistic, nuanced approach to risk.

Investment outcomes in superposition

In the real world, until the moment you cash out, every investment you make exists in multiple states at once. Think of an antique dealer looking at an old painting. Until the gavel falls, that painting isn't just one asset with one price. In their view, the painting represents several different investments. At one level, it can be a lost masterpiece, and there is enormous profit potential when it goes to a dedicated collector. You have the strikeout where the painting is identified as a forgery. All of these possibilities are true at the same time.

Now, our old way of thinking forces us to analyze these outcomes one after the other. We run the numbers for the best-case, the base-case, and the worst-case scenarios. However, this approach may not align with the perspective of an experienced dealer or a proficient investor. You're not considering these possibilities sequentially as an independent context; you're feeling their weight all at once. How does the risk of a forgery change what one would pay? How does a shift in the economy affect which collector might buy it? These factors form a complex network rather than a linear relationship, and our traditional models struggle to effectively handle this type of relationship.

So what happens when we stop forcing a single answer? That's what a quantum-inspired approach does. It frees you from having to select one story early on. Instead, it gives you the tools to simultaneously consider all those potential futures with all their complex connections. You can actually model how the doubt over authenticity impacts the optimistic price and how a hot auction market changes the downside risk, all simultaneously. Instead of predicting one outcome, you're mapping the entire probability landscape.

Of course, this mental oscillation can't last forever. The moment the dealer shakes hands on a price, all those shimmering possibilities vanish. The superposition collapses, leaving you with a single reality that could be either a triumphant payday or a painful lesson. That final deal is the "measurement" that makes everything real. Imagine how this idea gets even wilder if you apply it to a whole portfolio, where every investment is in its state of superposition, constantly influencing the others.

Portfolio superposition as integrated analysis

Let's scale up from a single investment to your entire portfolio. True diversification extends far beyond holding different assets to managing a web of interconnected potential outcomes. Picture a corporate treasury where cash is allocated across business units, currencies, political jurisdictions, and time zones. Each of those allocations isn't a fixed value, but a set of possibilities reflecting a superposition of gains, losses, and stable returns. Furthermore, these states constantly influence each other rather than remaining isolated.

Traditional theory relies on historical correlations, like how tech and healthcare stocks moved together in the past, or recollecting the old dance between emerging and developed markets. The fatal flaw here is the assumption that these relationships are static. It's like trying to navigate a living, changing environment using a static map from last year. This approach effectively "collapses" the dynamic, fluid nature of markets into a simplistic and often outdated pattern.

A quantum-inspired approach keeps these relationships in superposition. It acknowledges that their future state is uncertain and interdependent. Take a rate change announcement by a central bank. A classical model might predict a single, linear outcome. A quantum-informed view sees it generating a spectrum of potential reactions that ripple through the entire financial network at once. The impact on tech stocks, real estate, currency values, and corporate debt isn't predetermined. In quantum thinking, it exists as a probability distribution of outcomes.

Now think of running a diversified conglomerate. A savvy CEO views business units as interconnected rather than isolated silos that will react independently based on last quarter's report. They know a supply-chain shock in one division will impact manufacturing in another, which will alter the marketing strategy for a third. They see the cascading effects in real time.

In this light, the efficient frontier emerges as more than simple historical pairings. It comes from understanding how these superposed probabilities interact. The goal is to find asset combinations that provide complementary benefits across a range of futures, not just one. This means managers can maintain this flexible, probabilistic view of the portfolio until market events such as an earnings call, a geopolitical event, or a trade deal force those possibilities to resolve into a single reality. Power lies in the ability to retain optionality until the system finds no other path and collapses.

Decision-making under uncertainty

Classical decision-making is rule-based. To understand how it works, imagine a series of gates. We process each piece of information through an "if yes, do this" or "if not, do that" gate. The right path is determined by our end goal. This goal dictates how we should react to each new data point and whether we proceed to the next gate or stop.

This system faces two big problems. The first is not knowing how the gates connect to our objective. Here, decisions become ad hoc or just a matter of chance. The second and more profound challenge is when the information does not provide one clear answer. Instead, it provides us a range of possibilities. The only way to discover which one is real is to act. Your action itself collapses the range of possibilities into a single result. You then proceed from that new point. This is decision-making under true uncertainty. We typically use statistical models to try and squash this uncertainty into a single most likely value.

Quantum theory introduces a radically different kind of uncertainty. Here, uncertainty is not a flaw in our measurement; instead, it is a fundamental feature of reality. Heisenberg's uncertainty principle shows us that a particle does not have a precise position and momentum at the same time. These properties are genuinely undefined until the act of measurement forces the system to choose a value. Most financial uncertainty stems from our incomplete knowledge. In theory, with perfect information on every market participant, we could predict outcomes accurately, resolving uncertainty. The issue is that it is impossible to gather and process that information, regardless of its existence. Additionally, gathering and processing information alter the original data.

However, some parts of the financial system may indeed exhibit fundamental uncertainty. These are situations where multiple potential outcomes truly coexist until a market event, like a major trade or earnings report, collapses them into a single reality. This implies that for some market phenomena, more information will never provide a perfect answer.

To state it clearly, uncertainty is not just a lack of information. It is a basic feature of reality. The decision-making process is therefore a complicated combination of potential outcomes that may not follow classical rules.

Quantum foundations for financial analysis

Before we apply quantum ideas to finance, we need a quick guide to the core concepts themselves. Think of this as your quantum toolkit.

Superposition means a quantum system exists in all its states at once. Each state has a probability amplitude, unlike classical probability, where outcomes are separate. Superposition allows multiple states to coexist until a measurement forces the system to collapse into one. In finance, an asset's value exists in a superposition of prices. This continues until a trade measures it and collapses the possibilities into a single market price.

Entanglement occurs when particles become linked. No matter how separated they are, measuring one instantly affects the other. In markets, entangled assets behave similarly. Trading one security can instantly influence a correlated security in a way that feels like it goes beyond normal cause and effect.

This interference effect is crucial for understanding market anomalies. In classical thinking, if you have two positive factors affecting a stock, you'd expect the combined effect to be simply additive. But in quantum-like market interference, these factors can actually cancel each other out or create amplified effects that seem disproportionate to their individual impacts. In finance, this is when multiple market forces combine to create a result that is far different from a simple sum of their parts. Think of individual trading algorithms interacting to create unexpected market-wide effects.

Crucially, this collapse is instantaneous and irreversible. For example, once a trade executes at Q$50, all other potential prices (Q$49, Q$51, or Q$52) are immediately eliminated, not just temporarily suppressed. The market has "chosen" one reality from the superposition, and there's no going back to the unmeasured state. This is why post-trade analysis often seems obvious in hindsight, even though the outcome was genuinely uncertain beforehand. This action collapses all potential market states into one realized outcome and often irreversibly changes the future possibilities of the system.

Embracing the flux: finance as a dynamic system

As we've seen, finance is the machinery of wealth. It creates it, moves it, and tracks it. But we must remember this entire system is built on a foundation of constant, unpredictable change. Dynamic forces exert pressure on every aspect of the system – from the largest players to the most intricate processes. We know some of them, such as economic data, tech innovation, and political drama. But others are completely unknown. We only see them when they erupt, like when a Ming dynasty "chicken cup" sells for $36 million at auction. These unknown factors can rewrite market rules overnight, turning certainty into guesswork and solid valuations into speculation.

This relentless change forces us to build decision-making frameworks that can handle fluctuation and manage true uncertainty. The issue is that most of our current models assume the world is static. They treat uncertainty as a set of odds for outcomes, a variable to be calculated. What we need is to see uncertainty as a regular and fundamental feature of the system. This does not mean we throw out accounting and our accumulated wisdom. Our financial reports are still crucial. But we must change how we analyze that data to make decisions for a future that is fundamentally unknowable.

The ever-changing financial landscape

Finance has never stood still. An intricate mix of technology, geopolitics, and raw human behavior has driven its evolution since the Middle Ages. What started with barter evolved into a complex global network, thanks to innovations such as banking, stock exchanges, and digital tech. And to be honest, the barter system was probably just as complex. We just didn't have the tools to analyze how it worked.

Look at the Italian merchant banks in Florence, Venice, and Genoa from the 12th to 14th centuries. They changed everything with one simple development: credit instruments. Families like the Bardi and Peruzzi of Florence built vast banking networks across Europe. The Peruzzi had branches from London to Cyprus by 1325. But their story also teaches us about risk. Both families went bankrupt in the 1340s when King Edward III of England defaulted on his loans. Sound familiar? Perhaps that was our first lesson in sovereign risk and a tacit acceptance that there is no such thing as a risk-free rate.

Next came the need to record these transfers. Those same bankers perfected double-entry bookkeeping for their time. It had been practiced for ages but was finally documented by Luca Pacioli in 1494. They created the first real bank ledgers, the *libri segreti*, to track every debit and credit.

The pace of change never slowed. Each crisis spawned new systems. The 17th-century Dutch revolution gave us the first modern stock exchange. Then Dutch Tulip Mania in 1637 showed how crowd psychology could inflate a market to absurdity, a lesson we have failed to learn repeatedly. The 1720 South Sea Bubble in England demonstrated how speculation could create and destroy empires of wealth overnight, leading to the first major market regulations.

The 20th century accelerated chaos. The 1929 crash revealed systemic risk. The Bretton Woods system collapse in 1971 ushered in the wild era of floating currencies, creating new opportunities for financial gain and loss. Computerized trading in the 1980s changed market dynamics forever. This change had negative consequences for us on Black Monday in 1987, when automated systems amplified the market crash.

The 21st century arrived swiftly. The 2008 crisis served as a lesson in financial entanglement. The value of a house in Florida instantly diminished the worth of derivatives in Berlin, for reasons that few people understood. Blockchain emerged, challenging our very definition of money with cryptocurrencies. Artificial intelligence (AI) supercharged algorithmic trading, and social media, with no fund invested, emerged powerful enough to move markets as emphatically as any fund.

Through all this history, the ways we create and transfer wealth have transformed radically. We are now seeing the first use of quantum techniques in markets. Yet, through it all, financial reporting has barely kept pace. The reports themselves have changed, but they still force us to view a dynamic, quantum reality through a static, classical lens. We still use these snapshots of the past to navigate an uncertain future. Even worse, we treat these figures as absolute truth and need to move past this intellectual failure. We must align our thinking with the fundamental reality that uncertainty is not a problem to be solved, but the very environment we operate in.

Dynamic versus static strategies

We have two contrasting mindsets to deal with. Static strategies operate from a classical view and assume financial states are fixed and isolated. Dynamic strategies, however, embrace a more quantum view. They acknowledge that multiple potential states can exist at once, as in superposition, and that outcomes are deeply interconnected, as in entanglement.

Static strategies rely on rigid, long-term plans. They set fixed goals and timelines, prioritizing predictability above everything else. This approach works perfectly in a stable world but fails catastrophically when the unexpected happens. We often blame the model for failing to predict a crash, overlooking the fact that it was never designed to do so. It was built for a future world that it believed would replicate the present but, in reality, doesn't exist.

Dynamic strategies are defined by their flexibility. They are built to evolve using real-time information, allowing for quick adjustments to new data or market shocks. They constantly reassess plans to align with current reality, not outdated assumptions. You will have to remain in a constant state of alertness to capture the signals of change.

Now, we should note a crucial distinction. This business dynamism is not exactly a quantum state. It is about reacting to the latest information about a change that has already occurred. In a true quantum framework, all those potential states were always present, just with different probabilities.

Implementing these strategies requires new models. Take a simple example of a current ratio of 1.76. Our old habit is to compare it to the past, while a quantum view sees this number differently. Measurement collapsed a vast number of possibilities that the current ratio faced into a single value. That number emerged from a superposition of states such as unfinished production cycles, inventory in flux, and evolving credit terms, among others. This single metric is a snapshot of a probability wave that describes all future possibilities for that company, not just one. Understanding this allows us to prepare for a range of outcomes, not just one.

Market forces will leave behind and eliminate those who cannot adapt to this wave-like nature of finance. If you cannot play this new game, you might be better off not playing at all.

Historical context

History is littered with the wreckage of static strategies gathered across ages. When the first stock exchanges emerged in 17th-century Netherlands, the strategy was simple: buy for the long term and collect dividends. The slow pace of change and lack of information kept strategies static during the 19th-century Industrial Revolution. A small number of players could even shape the future to their will.

The static strategy of buying on margin helped create the conditions for the 1929 crash. This led to more sophisticated instruments and regulations. Modern automated trading systems should not deceive you. They may switch strategies, but they are still static, rule-based, and linear. This vulnerability was exploited on Black Monday in 1987, leading to the emergence of circuit breakers.

The IT revolution broke the old model of valuing tangible-asset businesses. Many failed to see that dot-com companies needed a new valuation framework. Although the 2000 bubble burst was painful, it taught us that we need to measure the critical value of intangible assets differently.

Few foresaw how a U.S. housing collapse could trigger a global contagion in 2008. Complex financial products entangled the world's economies, proving that measuring one market instantly affects all others. Regulators, wiser now, stepped in with new rules.

Every one of these crises is a lesson in quantum finance. Market bubbles are pure superposition collapse. Asset values can fluctuate between multiple states until a crisis triggers a sudden drop to a new price. Our interconnected system creates entanglement, where a problem in one area instantly becomes a problem everywhere. Most importantly, these events show that interference and individual rational decisions combine to create a collectively irrational outcome, something no classical model could ever conceive.

Are you able to observe the quantum signatures now? Can you observe the superpositions, the entanglement, and the interference? Beyond being merely theoretical, this represents the hidden pattern of financial history. Our journey now is to learn how to integrate this quantum theory into our decision-making for a better future.

Looking ahead

The shift is fundamental. We must move from asking, "What is the current state?" to "What are all potential states, and how do they influence each other?" This transformation from static analysis to dynamic possibility-thinking opens entirely new approaches to risk management, portfolio optimization, and strategic planning. Yet embracing this quantum view creates its own paradox: the very act of trying to measure and understand financial markets begins to change them. In our next chapter, we'll explore this quantum paradox in depth, examining how the wave-particle duality that defines quantum physics also governs financial markets and why your choice of measurement method literally creates the reality you observe. Prepare to discover why two expert analysts can arrive at wildly different valuations for the same company and why they're both right.

Chapter 2

The quantum paradox in finance

"*The seagull dives where fishes and sharp stones coexist*," goes an old Quantaland proverb.

Finance director Andrew released a glowing report on a new product line and expected a boost in stock prices. He was shocked to see prices plummet instead as analysts' scrutiny sparked unexpected sell-offs. His report reshaped the market he aimed to predict. This chapter explores why markets act like quantum particles, shifting with observation. You'll discover how wave-particle duality explains varying financial data perspectives, recognize the uncertainty principle's impact on decisions, and explore examples from Quantaland and beyond, showing how measurement affects outcomes, why accuracy and timeliness can't coexist, and how financial statements exist in multiple interpretations simultaneously.

Beyond analogy: when finance acts quantum

In the previous chapter, we established a groundbreaking concept: finance operates similarly to the quantum world. Both realms are defined by inherent unpredictability, where the act of looking changes what you see, and where multiple potential realities coexist until an action forces one into existence. Now, this doesn't mean atoms and assets are the same. Quantum principles often provide useful descriptions of how financial systems operate.

Take the price of a stock. Before a major event like an earnings report, people speculate about the number. But the true value isn't a secret number waiting to be discovered. It exists in a superposition of all possible prices. The earnings announcement acts like opening Schrödinger's box, collapsing all those possibilities into the single reality of the opening trade price.

This "observer effect" is everywhere. The moment a hedge fund's massive position becomes public knowledge, it moves the market. The observation itself changes the outcome. Additionally, similar to the spooky phenomenon of quantum entanglement, financial markets exhibit strange

DOI: 10.4324/9781003619635-2

interconnections. Remember 2008? Assets that were supposed to be great for diversification – stocks, oil, and even gold – all crashed together. They became entangled in a panic for cash, proving that no asset is an island. The old guards, trained in classical models of correlation, were completely blindsided.

This points to a deeper truth. Quantum mechanics tells us reality is probabilistic at its core. Traditional finance, however, is built on a deterministic idea that with perfect information, we could predict the future. We ignore the simple fact that markets are probabilistic beasts. No outcome is ever 100% certain. The possibility of being wrong is not a flaw in our data; it is a fundamental feature of the system.

Some researchers are now using quantum computational models to explain how markets really work. However, our entire system of financial management and reporting still largely ignores these quantum properties. Our goal isn't to throw out our old models but to learn how to implement and interpret them through a quantum lens.

The implications are profound. If markets truly operate according to quantum-like principles, then perfect prediction would be impossible in principle, not just in practice. Presently, we seek to eliminate uncertainty or at least its adverse impact. If uncertainty is fundamental, then we will have to change our thought process and learn to live with it. As you explore quantum finance further, remember that these aren't just academic curiosities. They represent potentially powerful new ways to understand the world of finance, where we spend our professional lives. This will provide us with a different view of the management principles and performance metrics.

The dual nature of financial reality

Let's look at one of quantum theory's weirdest ideas, wave-particle duality. Light acts as a spread-out wave in some experiments and a discrete particle in others. What you see depends entirely on how you choose to look. Financial assets do the exact same thing. To a long-term investor looking at an annual chart, a stock price is a smooth, wave-like trend. It has momentum, support levels, and resistance, making up the classic tools of technical analysis. But to a high-frequency trader watching tick-by-tick data, that same stock is a series of discrete, particle-like jumps. The price doesn't glide smoothly from Q$100 to Q$101. The price "quantum leaps" between Q$100 and Q$101 in increments determined by the minimum tick size.

This isn't just a neat analogy. It explains how two entirely different types of traders can both profit from the same assets. They are not seeing different things; they are observing different behaviors of the same underlying reality – just like physicists deciding to measure light as either a wave or a particle.

Wave-particle duality in financial reporting

Now let's apply this wave-particle duality to financial reporting itself. Your financial statements can be read in two ways. When you study trends and patterns over time through cash-flow statements, income statements, or horizontal analysis, you are observing the wave-like flow of wealth creation. But when you zoom in on a specific journal entry, an invoice, or a balance sheet snapshot, you are looking at the discrete particle of a single transaction.

Both views are based on the same economic reality. The individual transactions, behaving as particles, combine to form the waves that represent long-term trends. However, each perspective alone does not provide a complete understanding. An auditor digging into entries sees an entirely different picture than an analyst charting five-year growth, even though they are looking at the same company records. Your choice of lenses determines what you see. So before you dive into the numbers, ask yourself a simple question. Am I here to study the wave, or the particles, or both?

Corporate valuation duality – wave or particle?

Valuing a company presents us with the same duality. You can choose different methods. A discounted cash flow (DCF) model, or even perpetual cash flow, views the company as a wave of future earnings. But the moment you calculate the net present value, you collapse that future wave into a single particle, a number frozen in time. On the other hand, an asset-based valuation, or a liquidation value, is a pure particle view. It is a snapshot of value right now, ignoring future potential.

Doesn't it seem obvious now? Your choice of valuation method dictates what you will see. It is no wonder two excellent valuers can arrive at wildly different numbers for the same company. They do not measure different things. They are examining the same subject from various perspectives. Therefore, the next time you see a valuation dispute, do not argue about the number. First, agree on the method of observation. A buyer who wants to run the business will value the wave. A corporate raider looking to sell off assets will value the particle.

Duality in revenue recognition: an event or flow?

Revenue recognition rules, such as IFRS 15 and ASC 606, formalize this quantum duality. They demonstrate that we can perceive revenue as either a particle or a wave. There are two core patterns. Point-in-time recognition sees revenue as a particle, a discrete event when a product is delivered or a sale is completed. When we recognize revenue across time, it appears as a wave. You can see this wave in the continuous flow of value from a subscription service or a long-term contract.

Take a two-year mobile phone contract paid monthly. Is it 24 separate particle-like transactions? Or is it one continuous, wave-like service relationship? The answer is both. It depends on why you are measuring. When future usage is uncertain, the wave-like nature becomes even more pronounced.

How we choose to measure directly changes what we see. Discounting all future subscription payments to their present value transforms the wave-like nature of revenue into a particle-like representation. Recognizing revenue each month as service is delivered preserves the wave. Neither is more accurate than the other. They are complementary views of the same economic reality, just like light's dual nature. The real problem is that we often apply these accounting standards on autopilot, without appreciating this deeper duality. This blind spot can leave us with a serious lack of perspective when analyzing a company's true performance.

Capital structure duality

This wave-particle duality is perfectly visible in a company's capital structure. The ongoing quest for an optimal mix of debt and equity behaves like a continuous wave. In contrast, the actual financing decisions that a company makes, such as issuing a specific bond or selling stock, are discrete particles.

The wave view is all about fluid, continuous optimization. A company is always subtly adjusting its leverage target, swayed by market conditions, cash-flow stability, and growth plans. Think of a Quantaland firm slowly tuning its debt ratio over quarters. This is not a search for one perfect number. It is a navigation of a range of possibilities, much like a quantum wave function. The ideal structure is a moving target, shifting with changes in interest rates, tax policy, or dividend demands. Rigid models like Modigliani-Miller, which assume perfect markets, completely miss this dynamic, wave-like reality.

Next, we explore the particle view, which represents the world of concrete action. The decision made by Quantaland Engineers to launch a Q$50 million bond is an example of a particle. It has specific traits – an amount, an interest rate, and a maturity date. Either you initiate the deal, or you don't. There is no middle ground. Major events such as an initial public offering (IPO) represent a significant step forward. One day, the company is private; the next, it is public. The change is instant and absolute.

Your perspective dictates your strategy. When you are thinking long term, you work with the wave, making gradual adjustments. When you are executing a deal, you are dealing with particles, making binary choices. You need both views. You cannot set a strategy without understanding the wave-like spectrum of possibilities, and you cannot execute that strategy without the decisive particle-like actions that make it real.

Duality in cost behavior

Cost behavior also has its waves and particles. How you see your costs depends entirely on the reason you are looking into it.

From a strategic height, costs behave like waves. They show continuous, predictable patterns that change with business activity. Mixed costs are the perfect example. Take the maintenance budget for Quantaland Manufacturing. It has a fixed base cost and a variable part that changes with use. You can forecast its general flow, but the exact number for any given month exists within a range of possibilities. Even complex overhead costs reveal wave-like patterns over time, with seasonal peaks and efficiency drives creating interference – much like quantum waves.

But as you zoom in to the operational level, costs become particles. They are specific, measurable, and binary. Hiring a new supervisor for Q$60,000 a year is a cost particle. You make the hire, or you do not. The decision transforms the potential outcome into a definitive reality. Purchasing a new production line for Q$2.5 million represents a significant increase in both capacity and cost. It is not gradual. Each department, from marketing to R&D, acts as its cost particle with a defined budget. Specific items, such as software licenses or insurance premiums, also act as their own cost particles with defined budgets. You can turn them on or off.

Your perspective shapes what you see. During strategic planning, you must think in terms of waves, analyzing trends and probabilities. When approving a purchase order or a new hire, you are dealing with particles, making yes-or-no decisions on discrete costs. Mastering cost management requires that you know which lens to use and when. You need the wave view to set the strategy and the particle view to execute it.

The inescapable trade-offs: finance's uncertainty principle

Heisenberg's uncertainty principle reveals a fundamental truth. In the quantum world, you cannot know everything at once. The more precisely you measure a particle's position, the less you know about its momentum. This is not a flaw in our instruments but a feature of reality. Finance operates under the same constraints. The more precisely you focus on one metric, the blurrier another becomes. Furthermore, similar to physics, the process of measuring itself alters the outcome. Let us explore how this plays out in financial decision-making.

Accuracy versus timeliness in financial reporting

Every finance leader faces this core dilemma. You cannot have a report that is both perfectly accurate and immediately available. Precision requires time, and often we don't have it on our side.

Consider the monthly close at Quantaland Manufacturing. They can give you preliminary numbers right at the month's end. You get revenue estimates, rough expenses, and a ballpark profit figure. The information is timely enough for quick decisions, but it lacks precision. To get those audited, pristine numbers, you must wait. By the time you have that precision, the information is already a month old. You prioritize accuracy but sacrifice timeliness, resulting in January decisions based on December's reality.

You see this same trade-off between quarterly and annual reports. Quarterly figures are a good, timely snapshot. Annual reports are far more precise, but they describe a world that has already moved on, threatening to make the reported figures inconsequential.

Outside of strict rules, this choice also becomes an analysis of cost-benefit. You must ensure that the high cost of precision and the potential lost opportunity due to delay do not outweigh the benefits you receive from it.

Budgeting precision paradox

Now consider the budgeting process at Quantaland Engineering. Here lies another quantum trade-off. The more precise and detailed your budget is, the more rigid it becomes. Imagine crafting a perfect annual budget, nailing every single expense. It feels like a triumph of control. But that very precision locks your managers into the assumptions of a past reality. They will chase that specific sales target and might ignore a new, better opportunity. It is the financial equivalent of measuring a particle's position perfectly while having no idea where it is going next.

Rolling forecasts try to fix this. They sacrifice some details to gain responsiveness. But even they cannot escape the core trade-off between detail and adaptability.

How auditing changes what it measures

The audit is a wonderful example of the observer effect. Measuring compliance with an audit alters the behavior under observation.

If the inventory manager at Quantaland Manufacturing knows the auditors are coming, everything changes. The team starts counting stock more often, enforcing procedures perfectly, and keeping impeccable records. There is nothing wrong with this. But it means the audit captures a sanitized, best-behavior version of reality, not the normal day-to-day operation it intended to capture.

The cost of compliance itself is a measurement effect. Every dollar and every hour spent on auditing is a resource not directly spent on the growth of the business. This is not an argument against audits but a reminder that audits represent a trade-off. You are swapping operational efficiency for measurement precision. A more thorough audit costs more and disrupts more.

The effect even changes management decisions. Knowing a transaction will be hard to explain, you might avoid it altogether, even if it is a good deal for the company. You may choose simpler, less optimal paths to make your life easier with the auditors. The measurement changes the outcome.

Performance measurement distortions

Nowhere is the uncertainty principle clearer than in performance measurement. The second you choose a key performance indicator (KPI), you change the behavior you are trying to measure.

Please observe the sales team at Quantaland Engineering introducing a clear monthly revenue target and reviewing the results. The team will push for last-minute orders, negotiate deep discounts, and even strain customer relationships with pestering. The revenue will look great, but profitability may fall apart.

This is the uncertainty principle in action. You cannot have a perfect measurement of one metric and expect natural, balanced behavior across all goals. This is so well known that it has a name, Goodhart's Law, after economist Charles Goodhart. It simply says that when a measure becomes a target, it stops being beneficial. The act of measuring distorts the result.

Risk assessment paradox

One of the most dangerous trade-offs in uncertainty lives in risk management. The better you get at measuring the known risks, the worse you become at seeing the risks you do not know.

Please take into account the risk committee at Quantaland Bank. They pour immense effort into modeling credit risk, market risk, and operational risk. These are the classic measurable risks backed up by significant research. However, this focus tends to obscure the unknown. The 2008 crisis was not caused by a math error involving a known risk. It was caused by risks that existed outside all the popular models.

The uncertainty principle tells us that sophisticated risk modeling makes you vulnerable. You start to ignore anything you cannot put in your model.

This does not mean we should stop measuring, but it means we must be smarter about it. We must recognize these trade-offs and make conscious choices. Do we need precision, or do we need flexibility? Do we need accuracy, or do we need timeliness? A sophisticated finance professional embraces this uncertainty. They use it as a guide for when to measure tightly and when to let go, while never forgetting the core lessons. In finance, just like in physics, the act of looking changes the thing you are looking at.

Embracing the cloud of possibility

In the quantum world, particles do not have a definite state until you look at them. They exist as a cloud of probabilities. This is not a flaw; it is how reality works. Financial markets operate the same way. A stock's true value, a company's credit risk, and a project's outcome are not fixed numbers waiting to be found. They are probability distributions. When Quantaland Investments values a stock, they are not looking for one price. They are mapping a range of possible prices. That range only collapses into a single number when a trade happens or a regulatory report is filed. This indeterminacy is a core feature of finance, not a sign that our models are faulty. We must accept this reality when we make decisions. Sometimes this fuzziness arises out of a lack of information, but mostly it is fundamental.

Financial statement interpretation uncertainty

Here is a thought that might unsettle you. The same transaction can create different financial statements based on the accounting policy you choose. One economic reality can support multiple perfectly legal financial pictures.

Take Quantaland Manufacturing. They spend Q$2 million a year developing software. Their accountants could treat it as pure research and expense it all immediately. They could also call it development, capitalize it, and spread the cost over years. The useful life they choose is another variable. Both approaches obey the rules, but they show wildly different profits and asset values. So which one is the true picture?

The answer is both and neither. Your financial statements exist in the superposition of possible interpretations. It is the reader, the banker, and the investor who collapse them into a single reality through their analysis. Your banker might see a capitalized cost and think you are inflating your assets. When your investor sees an expensed item, they may assume you're being cautious. The same event creates different realities for different people.

This uncertainty goes beyond simple policies. It extends to many areas, including when to book revenue, how to classify a lease, when to write down goodwill, and so on. These are all judgment calls that collapse the probability wave into a specific reported number. This flexibility is often misinterpreted as an opportunity to manipulate. But, in practice, it is proof that financial reality is not fixed. It represents a range of possibilities that various observers will interpret and realize in different ways.

Valuation indeterminacy

Traditional finance encourages us to search for a single, true value for an asset. This is a myth. In reality, an asset's value is a probability distribution, not a point.

Look at Quantaland Investments' real estate portfolio. A DCF model might value it at Q$50 million. A market-based approach using multiples might say Q$55 million. The cost to replace it all might be Q$45 million. Which one is right? They all are. Each method captures a distinct aspect of value probability. DCF techniques estimate future income and convert it into present value. The market approach captures today's sentiment. Replacement cost gives a fundamental floor. Between all these values, we can see that the true value is not a number. It is the zone where these probabilities overlap. And each model itself is built on assumptions that have their ranges.

This is why fair value is such a powerful concept. It does not pretend to determine an objective price. Mark-to-market is not about discovering the correct price. It is about selecting one data point from a vast distribution of possible values. Prices swing wildly not because value changes by the second but because different traders are collapsing the probability wave at different moments. In a crisis, the range of possible values explodes, and prices can jump from one extreme to another. Every trade is just a temporary collapse of the superposition.

Credit risk assessment probabilities

Old-school credit analysis asks a simple question. Will this borrower default, yes or no? In contrast, quantum finance believes that this is the wrong question. Default is not a binary event. It is a probability wave that changes over time.

When the Bank of Quantaland looks at a loan, they are not making a "yes or no" prediction. They are building a probability distribution. What is the chance of defaulting in year one? Year two? Year five? This is why the time to default is so crucial. These probabilities are not fixed and change with every new piece of economic data.

This gets even more complex when you look at a whole portfolio. The risk of a loan portfolio is not just the sum of individual loan risks. Defaults can become correlated in unpredictable ways. In 2008, mortgages across the country defaulted together in a way no model had predicted. While we understood the individual risks well, we were completely unaware of the correlated, portfolio-level risk.

This means portfolio risk exists in a quantum superposition. It is a complex interaction of countless probability waves. A shift in the economy or a new regulation can collapse that entire superposition into a default reality that was invisible when we looked at each loan independently.

Project finance uncertainty

We are all taught to swear by the net present value (NPV). We treat that single number as gospel. But it is a lie, though useful, but a lie nonetheless. Every NPV is really a probability distribution pretending to be a point.

Quantaland Capital evaluates a Q$10 million project. The model spits out an NPV of Q$2.3 million. This number is not a fact but the center of a vast cloud of possibilities. That cloud is built on assumptions about cash flow, discount rates, terminal value, and taxes, among others. Each of those assumptions consists of a range of values. We just picked one number from each range to get our neat and seemingly accurate answer.

A simple sensitivity analysis shows the truth. Change the discount rate by 1%, and the NPV might swing from Q$1.8 to Q$2.8 million. Tweak the growth rate, and the range expands even wider. Tools like Monte Carlo simulation make this explicit. They run thousands of scenarios to show you the real output, a probability distribution. You might learn that there is a 30% chance you lose money and a 10% chance you make over Q$5 million.

Some say this makes things too complicated. It does not; instead, it makes the inherent complexity visible. A single-point NPV does not remove uncertainty. It hides it from view. Thus, when a project fails, we act surprised, even though the possibility of failure was always there in the numbers we chose to ignore.

The prediction paradox: demanding certainty in an uncertain world

We finance professionals face an impossible paradox. The financial world demands certainty from us. Investors want precise forecasts, boards want accurate budgets, regulators want definitive risk numbers, and so on. But we all know, deep down, that the future is fundamentally uncertain. There are always variables we cannot control and events we cannot see coming.

This creates a quantum-like dilemma. The more confidence we project in our predictions, the less reliable they often are. A detailed forecast creates a strong illusion of control, which leads us to overlook the inherent unpredictability that we should acknowledge. We are stuck between what our stakeholders want to hear and what we know to be true.

Overconfidence in models

Financial models are necessary, as they help us simplify chaos and make plans. But they are also hypnotic and make us overconfident. Just like a quantum particle exists in multiple states, financial reality is a spectrum of possibilities, not one predictable path. When we start treating our models as reality, we blind ourselves to this perceived truth. We begin to believe we can predict the future. Quantum thinking is the antidote, as it forces us to see models as useful tools, not absolute truths, and to always remember that the range of possible outcomes is the truth.

Financial planning model limitations

Consider the importance that a long-term strategic plan enjoys. However, it is a masterpiece of model overconfidence. It is filled with detailed projections and scenarios that create the illusion of mathematical certainty. Quantaland Manufacturing's plan might claim an exact 8% annual growth, a 3.2% annual cost increase, and a 15% market share in three years. However, are these figures genuine predictions or merely optimistic wishes disguised as numbers to create an illusion of confidence? The annual budget is even worse. The annual budget compels managers to act as though they already know the results for the upcoming year. We begin to believe that the marketing department will spend exactly Q$2.4 million and that sales will generate exactly Q$45 million in new deals. These numbers then become a script that everyone agrees to perform, forgetting it is just an estimate, if not a guess.

Our enterprise resource planning (ERP) systems make this problem permanent and self-deceiving. They bake these assumptions into the software itself. The system "knows" how long inventory will sit and when and how customers will pay. It then starts to run the business based on these frozen models, oblivious to the fluid reality outside.

The illusion of precision in accounting

Even our accounting standards are just models. The Generally Accepted Accounting Principles (GAAP) and the International Financial Reporting Standards (IFRS) are sophisticated attempts to standardize reality, but they too have built-in limitations. Look at lease accounting, where you create a "right-to-use asset" by discounting future lease payments. This gives you a precise number on the balance sheet. But that number is built on a mountain of assumptions about the future. Would it be possible to accurately predict the lease term? Can you accurately assess the probability of renewal options? The precision we trust is actually an illusion.

Revenue recognition has the same problem. When Quantaland Systems sells a software contract, the accounting model forces them to split the revenue over years. While this appears precise on paper, at what point does the customer truly realize the value? It might be entirely different. During a crisis, these fair-value models keep churning out numbers with authority, even when there is no real market to trade in. The model creates a reality, an illusion that does not exist.

Credit scoring and lending models

Credit scoring models can be dangerously misleading. They are treated as infallible, spitting out a precise number like a 12.3% chance of default. Nobody can doubt the math – it is perfect. The problem is that the past might have become irrelevant. These models are trained using historical data, so they assume the future will look like the past. But we can never know what the future holds; that is the charm of the future. This is not a flaw that we can correct. It is a fundamental uncertainty, much like in quantum physics.

Quantaland Bank's model was trained during good times. It never saw a real crisis coming. Naturally, the system could not predict a market crash because it lacked historical data on such events. A quantum-inspired approach would be different. It would not try to pick one future state for the loan. It would hold the loan's risk in a superposition of states – stable, stressed, and distressed, all at once. It would accept that the true outcome will only be revealed when the future collapses into the present.

Transfer pricing models

Transfer pricing is a game of make-believe where the actors dress as look-alikes. The models presume that a "comparable" transaction can be found to establish a price that is fair and reasonable. But think about Quantaland Instruments licensing its patented tech to its German subsidiary. How do you find a comparable license for something that is, by definition, unique? Patents are rewarded only when they are not like anything else present. You are making an unfair comparison.

Two technologies, despite their similarities, may find application in distinct markets with varying risks. The model creates an illusion of objective pricing for a deal that is utterly subjective and unique. How do you find a market comparable to a proprietary algorithm, a unique business process, or a specialized manufacturing technique? The model demands comparability where none exists, creating an illusion of objective pricing for subjective business arrangements.

Actuarial and pension models

Actuarial models represent the ultimate deception. They project pension obligations decades into the future with alleged terrifying precision. However, these figures rely heavily on assumptions about mortality, discount rates, and investment returns. Change the discount rate by just half a percent, and Quantaland Manufacturing's pension liability might swing by Q$50 million. Adjusting the mortality rate could result in an additional shift of Q$30 million. These are not inaccuracies about the model or assumptions. They are guesses about the unknowable.

The real danger is when we believe these numbers. Companies make massive funding and strategic decisions based on these educated guesses. They treat the model's output as a future fact. Suddenly, a significant liability materializes, leaving everyone unprepared.

Probability and risk

Traditional risk management is like making a shopping list. You identify the risk, categorize it, and assign it a score. Quantum finance demands a complete shift in perspective. It asks us to see risk not as discrete items, but as dynamic, interconnected probability fields. This approach does not eliminate uncertainty. Instead, it makes uncertainty visible and something we can work with. By focusing on probability distributions instead of simple yes/no outcomes, we can make stronger decisions that prepare us for a range of possible futures instead of just one value.

Enterprise risk-management framework

Traditional risk frameworks mostly treat risks as isolated items on a checklist. A quantum view sees them as interconnected probability fields that change in real time. Risk is in a constant state of superposition, with multiple potential outcomes whose odds shift with every decision and market event. Think of risks like ripples in a pond, constantly overlapping and changing. Old-school risk management tries to measure each ripple. Quantum finance looks at the whole, ever-changing wave pattern.

Quantaland Manufacturing's risk register lists "supply-chain disruption" as a fixed, medium-probability event. This is a fantasy, as the probability of a disruption is not a fixed number. It changes with the weather, politics, and supplier health. It is a live probability field.

A quantum approach understands that risks are entangled. A customer delaying payment is a credit risk. But that delay causes a cash-flow problem, which is a liquidity risk. That liquidity risk leads to a delay in making a tech investment, which poses a strategic risk. The strategic risk ultimately harms the company's competitive position. Unfortunately, traditional models primarily see these as separate. In reality, they are deeply connected probability fields that interfere with each other, creating outcomes that no single risk model could predict.

We need better tools than only complex matrices. We need live dashboards that show these probabilities shifting in real time based on leading indicators. This will turn our risk management from a static report into a dynamic early warning system.

Financial statement risk disclosure

Today's risk disclosures are often useless, ending up as long, legalistic lists of everything that might go wrong. Although these disclosures meet compliance requirements, they do not offer stakeholders valuable insights into the likelihood or severity of potential risks.

Please review Quantaland Software's disclosure regarding a lawsuit. It says that losses are possible but not estimable. A quantum-inspired approach would say, "We see a 60% chance of a settlement between Q$2.5 million and a 15% chance it may exceed Q$10 million." This is a range of probabilities that are updated with new data. They are not a single number. This might seem ambitious, but probability-based assessment is already at the heart of accounting standards. We just need to be brave enough to use it.

This could change annual reports. Instead of a generic warning about an "economic downturn," we could show how a mild, moderate, or severe recession would change the probability distributions for our revenue and cash flow. The goal is not false precision. It is to replace vague fear with informed probability, giving stakeholders something actually useful for their decisions.

Capital allocation under uncertainty

Traditional capital budgeting judges each project based on its own NPV. Quantum allocation views the options as a portfolio of projects with probabilistically linked outcomes, where flexibility carries a monetary value.

Imagine Quantaland Projects has three expansions to consider. The old method would rank them by standalone NPV. The quantum method would map how their probability distributions interact. Projects A and B might both thrive in a boom, so they are correlated. Project C might do well in a downturn, offering diversification. This changes everything about how you allocate capital.

This is where real options add power. The Q$500,000 allocated for market research is not merely an expense. It is the price of an option. It buys you the right, but not the obligation, to enter a new market if the data look promising. The standard NPV might be negative, but the option value of future flexibility could make it a brilliant investment.

The goal of capital allocation should be to optimize the probability distribution of returns across the entire portfolio and not to maximize a single project's point estimate. This starts with accepting that a project's return is a range of possibilities, not one number.

Financial ratio analysis

We often treat financial ratios like some divine order. A current ratio of 1.2 is "good." A debt-to-equity of 0.65 is "bad." These context-neutral specifications are a trap. Benchmarking becomes useful and intelligent when we compare probability distributions and not point estimates. Instead of saying "our ROE is 12%, beating the 10% average," we should say "our ROE distribution gives us 75% confidence we will outperform the industry." This probabilistic view kills false precision and provides a true sense of performance and risk.

Take Quantaland Systems' debt-to-equity ratio of 0.65. A traditional analyst would panic, while a quantum analyst would build a confidence interval around that value. They see that, based on seasonality, transaction timing, and measurement noise, the true ratio has a 90% chance of being between 0.58 and 0.72. The number is not a fact. It is a range that changes the entire conversation from one of alarm to one of informed understanding.

Cognitive dissonance

Just like a quantum particle can spin both ways at once, a finance professional's mind must hold conflicting ideas simultaneously. You have to trust your forecasts completely while also expecting them to be wrong. You exist in this mental superposition until a decision, a board vote, or a market crash forces one view to collapse into action.

This cognitive dissonance is not at all a weakness but a vital survival skill. Stakeholders demand that you achieve impossible, contradictory goals. Understanding this tension is the key to explaining why so many financial decisions look irrational from the outside. These decisions are not irrational; they are simply influenced by quantum factors.

Management reporting paradoxes

The CFO of Quantaland Systems constantly navigates a precarious situation. On one side, stakeholders demand a precise five-year forecast. On the other hand, they require the flexibility to change course quickly. These conflicting demands place the CFO in a quantum situation. You must fully believe in your numbers to present them confidently while simultaneously not believing them at all, which allows you to change course instantly.

This is not a flaw in the apparently contradictory request. Both needs are valid for different parts of the business. The CFO's job is to exist in this superposition. The solution is often linguistic with you labeling forecasts as "current best estimates" and not "future reality." This careful wording is a form of quantum measurement. It allows both certainty and uncertainty to coexist in the same sentence, balancing what stakeholders want to hear with what you know to be true.

Auditor independence conflicts

An auditor lives in two realities at the same time. They must be a skeptical prosecutor hunting for errors and a trusted consultant helping achieve the right answer. It is a quantum balancing act.

Please observe the auditors from Quantaland Systems as they review a complex revenue call. Their mind is in a superposition. One part is looking for problems while the other part is helping to solve them. For external auditors, this is even harder. They must be completely independent in their judgment, yet their business depends on a happy client. They are both independent and dependent at the same time.

This superposition usually collapses during a tough accounting debate, when one role must take over. The best auditors manage this by strictly separating the two modes. The technical judgment applies specifically to the workpapers. Relationship management is for the client meetings. Both statements are accurate, but they apply to different contexts.

Performance evaluation contradictions

Performance evaluation is a system designed to create dissonance. You are rewarded for hitting quarterly targets but promoted for building long-term value. Your bonus depends on this year's figures, but your career depends on the next five years. At Quantaland Retail, quarterly commissions push for aggressive sales that may ruin customer relationships. Come the annual review, and you are penalized for those ruined relationships. You end up being forced to choose between your bonus and your promotion.

A quantum approach would use probabilistic metrics that balance a short-term win against a range of long-term outcomes. Finance teams feel this, too. You are rewarded for hitting the budget, but you are also expected to challenge the same budget if it becomes outdated. You are supposed to execute the plan and question it at the same time.

Then there is the team-versus-individual conflict. Your incentive depends on your department's metrics. However, achieving company success may require collaboration that could negatively impact your department's metrics. Giving easy credit boosts sales but wrecks collections. You see why every professional is in quantum superposition, torn between optimizing for themselves and optimizing for the whole.

Financial reporting tensions

Quantaland Engineering faces an impossible choice every year. Their investors want details about their new manufacturing process to understand the competitive edge. But disclosing those details would be a gift to their competitors. They must be completely transparent with their customers while remaining entirely secretive about their strategies.

This is the core tension of financial reporting. You are balancing openness and secrecy like a quantum state. Accounting choices worsen it. Do you use conservative historical cost or volatile fair value? Regrettably, neither approach provides a complete picture. Fair value shows the current market mood but, at the same time, adds noise that hides the real business trend.

You cannot solve this tension and can only manage it. Quantum-inspired thinking helps, perhaps with probabilistic disclosures that give a sense of the range without giving away the secret formula.

Corporate finance decision biases

A mergers and acquisitions (M&A) deal beautifully demonstrates cognitive dissonance. The deal team brims with confidence about the synergies. At the exact same time, the risk team lists all the various ways the integration could fail and destroy those very synergies. Look at Quantaland Manufacturing's recent acquisition. The investment deck glowed with cultural fit and cost savings. At the same time, the risk assessment results warned of integration challenges and market shocks. Both the integration challenges and market shocks were accurate assessments, yet they had contradictory views of the future.

This dissonance is prevalent throughout corporate finance. You chase aggressive growth with a bold strategy, while your risk team enforces a conservative framework. You are supposed to be a daring gambler and a cautious protector at the same time.

Let us be clear. This is not a sign of a broken strategy. This tension is the heart of corporate governance. Every company must balance these opposing forces until a crisis forces a choice.

Living in financial quantum superposition

The goal of quantum finance is not eliminating these contradictions. The goal is to become comfortable living inside them. You must learn to hold multiple true perspectives at once until a decision forces a temporary collapse into action. Like a quantum particle, a finance professional must become at home in superposition.

This quantum view recognizes that most financial contradictions are not riddles to solve. There are tensions and contradictions to be managed skillfully every day. Your job is to develop the fluidity necessary to switch between these contradictory mindsets. You must know which hat to wear for which meeting, all the while remembering that you are, and always will be, wearing two hats at once.

Meanwhile, far from Quantaland

The rise and fall of WeWork (2019–2021)

The Quantum State: WeWork's valuation hovered in a quantum-like superposition. It was perceived as both a high-flying tech platform and a traditional real estate business until the IPO filing forced a market "measurement" revealing its true financial state.

The Story: WeWork's journey from a $47 billion valuation to near bankruptcy represents one of the most dramatic collapses in corporate history. WeWork was founded in 2010. Although WeWork operated in the real estate sector, it presented itself as a technology-driven "space-as-a-service" platform. This positioning yielded an extraordinary valuation growth, particularly after SoftBank's Vision Fund invested billions beginning in 2017.

In August 2019, WeWork filed its S-1 for an IPO, revealing alarming financial realities: the company had lost $1.9 billion in 2018 on revenue of $1.8 billion. The prospectus also exposed unusual corporate governance practices, including Neumann's nearly unchecked control and numerous conflicts of interest. One damaging revelation was that Neumann had personally trademarked the word "We" and charged the company $5.9 million to use it. This decision was later reversed after public outcry.

WeWork had been in a state like the state of quantum superposition. It was simultaneously worth $47 billion and nearly bankrupt. This duality persisted until the 2019 IPO filing acted like a quantum measurement, collapsing the narrative into a single reality of financial distress. Eventually, WeWork filed for bankruptcy in November 2023.

The Quantum Finance Lesson: WeWork's story is a clear illustration of how financial valuations can exist in several conflicting states at once. For an extended period, people perceived the company as both a cutting-edge technology platform and a typical real estate enterprise. These opposing perspectives persisted until the IPO process, serving as a moment of market observation that forced all possibilities to resolve into one undeniable reality. This event exposed the underlying uncertainty that was always present.

Heisenberg's uncertainty principle is also reflected in this dynamic: the more investors concentrated on WeWork's growth narrative, the less clarity they had regarding its financial fundamentals. When attention shifted to scrutinize those fundamentals, the appealing growth narrative quickly unraveled.

Practical Application: Investors in Quantaland should adopt a "quantum mindset" by simultaneously assessing a company's growth narrative and its core financials. For example, when valuing a firm like WeWork, cross-check optimistic projections against cash-flow metrics and governance red flags. Use tools like DCF analysis alongside qualitative due diligence to test both narratives. Recognize that public scrutiny, like an IPO filing, can collapse these possibilities into a stark reality, so stress-test valuations under pessimistic scenarios early.

Long-Term Capital Management (1998)

The Quantum State: Long-Term Capital Management (LTCM) relied on classical models assuming predictable market behavior, overlooking quantum-like extreme events that defied their probability calculations.

The Story: LTCM was founded in 1994 by John Meriwether, former head of bond trading at Salomon Brothers. Nobel Prize-winning economists Myron Scholes and Robert Merton joined as partners. LTCM used sophisticated models to make money by spotting pricing inefficiencies in similar assets. LTCM believed that markets would eventually behave rationally, and prices would revert to fair value. This worked well, and in its first three full years, the fund returned 21%, 43%, and 41% after fees. By 1998, LTCM had over $129 billion in assets against just $4.8 billion in capital.

LTCM's models underestimated the probability of extreme events. The normal distribution models used by LTCM suggested that a market movement of five standard deviations should occur once every 7,000 to 14,775 years.

In August 1998, Russia defaulted on its internal sovereign debt, triggering the kind of market shift that LTCM's models deemed virtually impossible. Yield of 10-year US Treasury bills dropped from 5.5% to 4.53% between August and October 1998. The S&P 500 fell by over 15% between July and October 1998. The Dow Jones Industrial Average declined by around 19% in just a few weeks. LTCM's equity declined by 44% over the month of August. Investors rushed to safer assets, and LTCM's convergence trades unwound violently under liquidity stress.

The Quantum Finance Lesson: LTCM's experience vividly demonstrates the shortcomings of financial models grounded solely in classical probability, which often fail to consider the "quantum-like" nature of real-world markets. In quantum physics, particles can overcome seemingly insurmountable barriers, a phenomenon that is also present in financial markets, where dramatic, seemingly impossible moves can and do occur, defying the predictions of traditional models.

The collapse of LTCM underscores a core insight from quantum finance: markets aren't always orderly or predictable. Rather than following smooth and continuous trends, they can undergo abrupt and unpredictable shifts – the quantum leaps that classical, statistically based models simply cannot foresee. By depending on risk-neutral pricing and similar models, LTCM created a false sense of precision and control, overlooking the profound uncertainties and nonlinear behaviors that are inherent in complex and interconnected financial systems.

Practical Application: Quantaland investors should embrace quantum-inspired risk management by stress-testing portfolios against extreme scenarios. These would include scenarios that classical models label as improbable. For example, run simulations assuming a Q$ market crash or sudden liquidity freeze like Russia's 1998 default. Use tools like value-at-risk (VaR) with tail-risk scenarios or Monte Carlo simulations with non-normal distributions to capture potential "quantum leaps" in market behavior. This prepares portfolios for rare but devastating events that classical models might miss.

Collapse of Enron (2001)

The Quantum State: Enron sustained a quantum-like superposition, appearing both highly profitable and deeply insolvent, until market scrutiny collapsed this duality into a stark reality of financial failure.

The Story: Founded in 1985, Enron initially operated as a traditional natural gas pipeline company. Under the leadership of CEO Kenneth Lay and later Jeffrey Skilling, Enron reinvented itself as an energy trading company and expanded into numerous markets, including electricity, broadband, water, and even weather derivatives. By 2000, Enron had become America's seventh largest company by revenue, reporting $101 billion in revenue and earnings. Enron's market capitalization peaked at approximately $70 billion.

Behind the story of success, however, lurked a different reality. Enron employed complex accounting schemes to hide debt, inflate profits, and maintain its investment-grade credit rating. Mark-to-market accounting and special purpose entities (SPEs) were two of the favored tools used. The company's chief financial officer, Andrew Fastow, created hundreds of SPEs and moved underperforming assets and debt off Enron's balance sheet there. Some of these SPEs even had names like "Jedi" and "Chewco," inspired by Star Wars characters.

Bethany McLean, in her March 5, 2001, Fortune article, "Is Enron Overpriced?", asked, "How exactly does Enron make its money?" This began the process that ultimately collapsed Enron's quantum state of simultaneous profitability and insolvency. The unraveling accelerated in 2001 when Skilling unexpectedly resigned as CEO in August. By October, Enron announced a $618 million quarterly loss and a $1.2 billion reduction in shareholder equity triggered by the unwinding of the SPEs. In December 2001, Enron filed for what was then the largest bankruptcy in U.S. history, with $63.4 billion in assets.

The Quantum Finance Lesson: Enron serves as a striking illustration of the quantum paradox in financial reporting. Through sophisticated accounting maneuvers, companies can project an image of existing in multiple financial states at once – profitable and insolvent – much like a quantum superposition. This illusion persists until external scrutiny collapses the duality and reveals the true financial condition, echoing the concept of Schrödinger's cat, where Enron's apparent success and underlying insolvency coexisted until closer examination forced clarity.

Enron's executives maintained a deep cognitive dissonance, allowing them to sustain conflicting financial realities simultaneously. This quantum-like illusion extended to the broader market, which temporarily accepted both the narrative of prosperity and the hidden insolvency. Ultimately, increased scrutiny forced the illusion to collapse, revealing the company's true financial state.

Practical Application: Quantaland investors should adopt a quantum-inspired view, recognizing that financial statements reflect a range of possible realities shaped by accounting choices. Scrutinize methods such as mark-to-market or off-balance-sheet entities for signs of manipulation. For instance, compare reported earnings against cash flow from operations to detect discrepancies. Treat complex financial structures, such as SPE, with skepticism, as they may hide a superposition of strength and weakness until market scrutiny forces clarity.

Looking ahead

The paradox is now clear: financial systems demand certainty while operating on fundamentally uncertain principles. We've seen how measurement changes outcomes, how the same reality can support multiple valid interpretations, and how the act of observation itself reshapes what we're trying to understand. But this quantum nature of finance reveals itself most dramatically in how wealth itself behaves. The values are less stable than balance sheets show; they're more elusive and transient. In our next chapter, we'll explore how wealth exists in quantum superposition, capable of sudden, dramatic leaps that can transform a company's entire financial reality overnight. Prepare to discover why that "solid" balance sheet might be more fragile than you ever imagined, and why the most dangerous assumption in finance is the belief that today's wealth will still exist tomorrow.

Chapter 3

Transitory states of wealth

"*The prey looks bigger as the eagle dives*," goes an old Quantaland proverb.

Elena, a seasoned investor, was pleased with the steady gains her diversified portfolio generated. Then a crypto market implosion reduced her wealth by a quarter overnight. This happened despite her cautious selection. Stability was a mirage, she concluded. This chapter shows how wealth exists in multiple states, like quantum particles, ready to shift suddenly. You'll discover how wealth behaves more like quantum particles than classical objects. You will examine real-world examples from the 2010 Flash Crash to cryptocurrency volatility and appreciate how accounting standards mask uncertainty. You will also explore industry-specific applications to transform your approach to risk and investment.

The myth of steady wealth

Wealth doesn't always move gradually; it quantum-leaps occasionally. Just like an electron can jump between energy levels without traveling the distance in between, a portfolio can lose a large part of its value overnight. A solid financial position can collapse without any warning at all. However, these aren't bizarre anomalies. They are the direct result of wealth existing in a quantum superposition of states, not in a single, fixed value.

Traditional accounting trains us to see wealth as a precise number on a balance sheet. The entire apparatus of audits, regulatory reviews, and financial analysis reinforces this illusion of a single, stable state. However, the wild fluctuations of the real world only become comprehensible when you embrace the quantum truth. That truth states that wealth is a probability distribution rather than a fixed value. This realization transforms your approach by shifting your goal from predicting a single correct future to preparing for a wide array of possible futures.

The surprise we feel during a market crash isn't caused by the change itself. It is caused by our own flawed understanding. We were looking for a particle and got hit by a wave. The next section will explore the specific

DOI: 10.4324/9781003619635-3

practices that create this illusion of stability and why dramatic change is often as real as the calm that preceded it.

Wealth in superposition: being rich and poor simultaneously

The old-school view sees wealth as a fixed number, changing slowly and predictably. Quantum finance shatters that illusion. It shows us that wealth exists in a superposition of potential outcomes, just like a quantum system. Though, in finance, it's not literally about particles, the analogy is powerful. Your net worth isn't a single number. It's a range of possibilities that only collapses into a definite value when an event, like a sale or a report, forces it to.

So what does this mean for you? Stop thinking in single points and start thinking in ranges. When you look at Quantaland Technologies stock, don't just pick one target price. Model the best case, the worst case, and everything in between. Use Monte Carlo simulations to map the entire landscape of possibility. This prepares you for any future, not just the one you hope for. This shifts your goal from futile prediction to intelligent probability management. You may already be using these tools, but with a classical mindset, you view them merely as a means to reduce your lack of knowledge. With a quantum mindset, you understand that all those states are simultaneously real. This will influence everything you do to create sustainable wealth.

Unrealized gains and losses as superposition states

Think of Quantaland Tech's stock right before earnings. Although it may be difficult to understand, the price does not reflect a single reality. It's a superposition of all outcomes: a 15% jump, a 10% drop, or no change. That Q$25 paper profit in your portfolio remains unchanged. It isn't real money but a quantum state of potential that only becomes real, or lost, when you sell.

This is why paper profits feel less real than cash in hand. They are indeed less real. They exist in a blur of futures that could vanish, double, or stay the same. Accounting's split between realized and unrealized gains is its feeble way of acknowledging this quantum truth. That Q$25 gain in market value is a probability, not a reality, until the trade settles it.

This superposition is even clearer in complex investments. Take Quantaland Manufacturing's emerging market bonds. A rising currency points to profit, but political risk warns of loss. The investment is both profitable and unprofitable at the same time. This state continues until a sale or crisis forces a definitive outcome. Our current accounting practices are not sophisticated enough to capture this true, simultaneous reality.

Fair-value accounting superposition

The purest expression of financial superposition is found in fair-value accounting. These assets exist as probability distributions and transform into a single reported number on the day of valuation. The most complex valuations, the ones based on internal models, are a perfect example. They hold multiple values at once, with each calculation method creating a different potential reality.

This exposes the enormous illusion of financial reporting, the illusion of precision. The Q$45 million property valuation should not be considered a fact. It is more likely the middle of a range of possibilities. Tweak a discount rate or a growth assumption, and you create an entirely new superposition of values.

Consider the portfolio of Quantaland Real Estate. Between quarterly valuations, those properties are in superposition. Could rents rise? Could cap rates change? Even the entire economy could shift. The quarterly mark-to-market act is the quantum measurement that forces all those possibilities to collapse onto the balance sheet as a single, deceptively precise number.

Contingent assets and liabilities

Contingent assets and liabilities are pure quantum states. They exist in a "maybe" condition until a specific event forces a resolution. Quantaland Systems' Q$20 million lawsuit is a perfect example. This liability both exists and does not exist. The potential outcomes, from a total win to a full loss, all coexist in a state of superposition. The court's final ruling is the measurement that collapses all these possibilities into a single financial reality.

Accounting standards try to handle this by asking us to pick the most probable outcome. But this is a classical solution to a quantum problem. It ignores the full spectrum of what is truly possible.

This principle applies to many areas. When Quantaland acquired the software company for Q$30 million plus an earnout, it effectively purchased a superposition of future costs and potential value. The earnout is both a future cost and a potential value. Its final amount will only be determined when future performance collapses the uncertainty.

Even tax audits are quantum events. The transfer pricing position for Quantaland Manufacturing's German subsidiary could lead to no change, a small penalty, or a massive tax bill. All these scenarios remain valid until the tax authority makes a decision that serves as the measurement.

The crucial takeaway is that if the outcome is different from what you reported, it is not an accounting failure. It was simply a different possibility from the superposition that became real. The potential for that outcome was always there. It's a different matter that you were unaware of its existence.

Going-concern versus liquidation values

Every company has two different values at the same time. There is value if it keeps operating, and there is value if it gets shut down and sold for parts. For Quantaland Retail, that's the difference between a Q$150 million and a Q$60 million valuation by two valuers. Both of these values are real. They exist in a quantum superposition until a decision, a restructuring, or a bankruptcy filing collapses it into one reality.

Management's choices in a crisis are not just business decisions. They are quantum measurements. The company's value collapses into one state or the other when it restructures or files for bankruptcy.

This duality explains the most confusing thing about stock markets: why a troubled company's share price makes no sense. It's because the market price isn't based on one value. It's the probability-weighted average of both the going-concern and the liquidation value. The market is pricing the superposition itself.

Our accounting system forces us to pretend that this superposition is nonexistent. We have to pick one value to report, usually the going-concern value. But in the boardroom and the market, both values are always present, and the tension between them is what defines how a company would act in distress.

Intangible asset valuations

The value of an intangible asset is one of the extreme exercises in quantum probability. A brand, a patent, or customer loyalty all lack fixed value. It has a range of values that change with every business move. Quantaland Consumer Products' brand is valued at Q$200 million today. However, that value is subject to constant fluctuations. A competitor's move could collapse it to Q$120 million. A successful new product could collapse it to Q$280 million. The brand simultaneously exists in all these states.

Goodwill impairment testing is our twice-yearly ritual where we try to collapse this superposition. One model says the goodwill is fine. Another model suggests it's impaired. The impairment decision is the moment we choose which reality to believe in, at least for now.

But the purest form of intangible superposition is the R&D project. Quantaland Pharmaceuticals' Q$50 million drug development program is both a brilliant future blockbuster and a complete waste of money. It is a quantum state of both immense and zero value. Only one event can collapse this superposition: the decision from the regulators. Until that day, the project, and the wealth it represents, exists in a ghost-like state of potential.

Foreign currency translation

The value of a foreign subsidiary is never a settled fact. Quantaland Manufacturing's European operation is worth €400 million, but its value in Q$ is a moving target. Until the quarter ends and we snap the exchange rate, the value exists in the superposition of all possible translated amounts.

This is why the "translation adjustment" to your equity is so jarring. Exchange rates don't cause a gradual change. They cause a phase change. One day, the subsidiary is worth Q$430 million; the next, it is Q$380 million. This adjustment reflects the accounting record of the superposition collapsing from one state to another, resulting in a sudden change in wealth that bypasses the income statement entirely.

Hedging is our attempt to control this quantum instability. A good hedge creates entanglement between the value of the derivative and the underlying asset. This keeps the overall financial superposition stable, protecting the balance sheet from currency volatility. However, if the hedge proves ineffective, the protective entanglement will break. The superposition collapses, and the full, violent swing of currency risk hits the books all at once.

When markets quantum-leap

Financial markets don't always move smoothly. Occasionally, they jump. Like quantum particles leaping between energy levels without passing through the space in between, market values can undergo sudden, discontinuous changes. These quantum jumps shatter the classical illusion of orderly, gradual price adjustments. Here's how this plays out in several critical areas.

Credit-rating fluctuations

Credit ratings are a splendid example of quantum finance. A bond from Quantaland Corporation can't slowly drift from a BBB+ to a BBB rating. It must transform discontinuously from one discrete state to another. This isn't a gentle slope; it's a cliff edge. The moment the rating drops, borrowing costs instantly jump higher, and debt covenants are breached. Instead of a gradual decline, it's a sudden phase change imposed by the rating agency's measurement.

This state transition triggers another one in the bond's price. The market price, instead of gradually adjusting to the news, instantly collapses to a new, lower level, creating massive volatility. Institutional investors, mandated to hold only "investment-grade" assets, find themselves compelled to sell all at once. This creates a selling avalanche that drives the price down far more than the fundamental news would suggest, reflecting a case of pure quantum amplification effect.

Earnings surprise effects

An earnings announcement is the market's version of opening Schrödinger's box. Before the news, a stock like Quantaland Software exists in a superposition of prices around Q$80, based on expectations of an earnings announcement of Q$2.50 per share. The announcement of Q$2.60 is the measurement that collapses that superposition. From a classical perspective, a 10-cent rise may not justify an 8% price jump. But in quantum terms, it makes perfect sense. The measurement itself amplifies the result, as all the probabilities collapse at once.

This amplified reaction is due to quantum interference. Every investor's expectation is like a probability wave. When the actual number hits, these waves don't just add up. They interfere. They can amplify each other (constructive interference), creating a massive price surge, or cancel each other out (destructive interference), causing a crash. Both extremes are far beyond what simple math models would predict.

The impact rolls over to future periods. Management's future guidance creates a new superposition for the next quarter far beyond setting a target. It establishes a fresh set of probabilities that the market will hold in its mind until the next earnings measurement collapses them into a new reality.

Regulatory approval processes

Regulatory decisions are a financial guillotine. They don't gradually change value; instead, they decapitate one reality and instantly replace it with another. And because companies in a sector are entangled, one decision can suddenly transform an entire industry. Events like a drug approval, a merger clearance, and a license grant don't create ripples. They create instant, sector-wide phase changes.

For Quantaland Pharmaceuticals, awaiting regulatory approval for a cancer drug, there is no middle ground. The company is in the superposition of being worth Q$2 billion or Q$200 million. The regulator's decision will collapse this superposition in an instant. There is no "kind of approved."

This binary nature collapses strategic superpositions, too. Quantaland Energy's plans for its offshore drill sites are in a state of possibility until the government decides on the permit application. The yes-or-no changes the budget line and proceeds to force a radical change instantly in strategy, redirecting billions from traditional exploration into alternative energy overnight.

Customer concentration risks

A major contract renewal is a quantum event. It bypasses a gradual adjustment in revenue expectations. It instantly collapses a company's future into one of two starkly different realities: growth or decline. This immediate leap changes everything about working capital needs and investment capacity right away.

Quantaland Services lives in this reality. With 60% of its revenue being tied to just three customers, each relationship exists in a superposition of renewal and loss right up until the moment the contract is signed.

This creates a form of financial entanglement. If one of these major customers hits financial distress, the effect isn't gradual. It's immediate and violent. Orders don't gently decline; they stop. Payment schedules shatter. The financial distress of the customer transforms unannounced into a working capital crisis for Quantaland Services, proving their financial states were never independent. They were deeply, fundamentally connected, a possibility we never explored.

The dangerous illusion of stability

Financial markets are great illusionists. Their calmest periods create the most dangerous fantasy, the fantasy of stability. This tranquility masks the underlying quantum uncertainty, where risks exist in a volatile superposition. Conventional models function flawlessly in these tranquil conditions, leading to a dangerous sense of complacency. We will now reveal how this apparent stability conceals fundamental uncertainties that can, and often do, abruptly collapse into crisis.

The abnormal normal distribution

Classical finance tries to force the messy, chaotic reality of markets into neat mathematical boxes. Iconic models like Black-Scholes are built on a foundation of elegant theory. But that theory rests on two flawed assumptions: that returns follow a normal distribution and that volatility is constant. Quantum market behavior shatters both of these ideas every day.

Normal distribution assumptions in corporate finance

We build our corporate finance tools on a flawed foundation. Capital budgeting models, like the NPV calculations used at Quantaland Projects, assume investment returns follow a pleasing, predictable, normal distribution. They might tell you a catastrophic loss is a once-in-50-year event. But if you've lived through three of them in the last decade, the model isn't simply wrong; it's dangerously misleading. The modeler will dismiss these as outliers. But in a quantum world, these "outliers" are the reality.

The fundamental error is assuming the future will be a statistical extension of the past. Quantum finance shows us that market returns exist in the superposition of potential outcomes. Our models mistake a temporary period of calm for a permanent law of nature. They assume gradual transitions, yet markets can abruptly shift from stability to crisis without any prior warning, resembling a quantum leap.

This illusion of precision infects everything. Look at how the Quantaland Project calculates its cost of capital (WACC). It uses a five-year average risk premium, pretending volatility is constant. In reality, volatility itself is in a quantum superposition, fluctuating between low and high states. Using an average is like preparing for both a calm lake and a perfect storm by packing just a raincoat.

Risk-management model failures

VaR models serve as monuments to this false confidence. They consistently underestimate tail risk because they assume market behavior is normal. They treat crises as statistical outliers, but in quantum terms, a crisis is a leap into an entirely different market state. This is a new reality where all assets move together, and liquidity vanishes, but we have not modeled for that.

It is no surprise then that Quantaland Bank's VaR model can confidently predict a maximum daily loss of Q$5 million, while the trading floor has actually lost Q$15 million multiple times. The model is not actually measuring risk; it is truly measuring the illusion of safety.

The failure extends to operational risk. Models based on historical loss data are useless against quantum operational failures. A cyberattack, a supply-chain collapse, or a regulatory earthquake doesn't have historical patterns. These events represent sudden shifts into a state of crisis that current models are unable to predict.

Financial and reporting model limitations

Across finance and accounting, our models have failed because they reject quantum reality. Let's quickly survey the wreckage.

- *Depreciation Schedules*: Quantaland Manufacturing assumes a machine will last ten years. Suddenly an innovative technology emerges, making

it obsolete in two years. This isn't gradual wear and tear; it's quantum obsolescence.
- *Allowance for Doubtful Accounts*: Your provision based on last year's default rate is meaningless when the entire economy jumps into a state of recession.
- *Pension Fund Actuarial Models*: They use long-term historical averages for mortality and returns. But these assumptions exist in the superposition of future possibilities, including developments in healthcare, that history cannot predict.
- *Transfer Pricing*: The "arm's length" price is based on stable market comparables. This is a fantasy when markets leap into a crisis and the comparables vanish.
- *Budgeting and Forecasting*: Assuming 5% revenue growth because that's the three-year average is a prayer or a hope but not a plan. It ignores the potential for quantum disruption.
- *Strategic Planning*: The five-year plan is the ultimate act of classical arrogance. It assumes gradual evolution when change happens through quantum jumps.

The core failure is always the same. We treat the uncertain future as a simple extension of the stable past, and we dismiss extreme events as outliers. Quantum finance teaches us that the future exists in a superposition of states. When the future collapses into reality, it often looks nothing like the past on which we based our models.

What lurks beneath the calm

We have discussed earlier why extended periods of market calm are deceptive. We know that they create the most dangerous illusion of all, the illusion of safety. This tranquility masks quantum risks that exist in a volatile superposition, waiting to collapse into crisis. Organizations become complacent, mistaking a temporary, stable quantum state for a permanent condition. They fail to see that stability itself is often the precursor to a dramatic phase change.

Complacency in internal controls

In stable times, companies shift their focus from defense to offense, from risk management to efficiency. This is when fraud risk actually increases. Management assumes controls are working and may even cut spending on them, unaware that risks are building up in the shadows. It's a false sense of security like a quantum system that appears stable but holds multiple potential states.

At Quantaland Systems, the internal audit team starts ignoring "stable" processes. This creates quantum blind spots where controls slowly degrade undetected. Every process control exists in a superposition between effective and broken. We only discover the truth when a crisis acts as a measurement event, collapsing the superposition and revealing the accumulated failures.

Liquidity illusion in corporate finance

Corporate treasuries fall into a liquidity trap during calm periods. They begin to believe market liquidity is constant. Quantaland Systems might be tempted to run with minimal cash, relying on credit lines that have always been there. This is a classic quantum illusion.

Those credit facilities exist in the superposition of being both available and unavailable. They only collapse into one state when you actually need to draw on them during a crisis. And that is precisely when banks suddenly tighten standards. Your lifeline can vanish at the moment you need it most.

Short-term funding is even more vulnerable. The entire market for commercial paper or short-term loans can quantum-freeze overnight. These funding sources are in a superposition of reliability. You only discover their true state when you try to refinance the old bonds, only to find there are no buyers for your new papers. The measurement itself leads to collapse.

Operational leverage risks

In calm markets, operational leverage can cause silent destruction. It turns small drops in revenue into catastrophic profit collapses. A shock change in demand, up or down, creates a wildly disproportionate impact on the bottom line.

During stable times, a high fixed cost structure looks brilliantly efficient. Quantaland Manufacturing locked in this structure to maximize profit above its break-even point. But this creates a quantum vulnerability. When demand suddenly falls, that fixed cost structure cannot adjust. You can't quantum-leap your rent or your payroll down. The very efficiency that drove profits now accelerates losses.

A company's cost structure exists in a superposition between efficient and flexible. Stable times hide this tension. A sudden demand shift is the measurement that forces a brutal collapse into one state, often triggering painful restructuring.

Customer and supplier concentration

Dependencies are invisible in good times. Quantaland Retail's reliance on a few big customers seems like a strength. Their credit ratings look solid, masking an entire range of outcomes. An economic shift acts like a quantum measurement. It collapses this superposition, instantly revealing whether those customers will pay on time or stall. You're forced to react to a new reality you didn't see coming.

The same is true for suppliers. Your relationship with a single-source supplier exists in the superposition of reliability and vulnerability. You only discover the true state during a supply-chain shock when you desperately need alternatives. Unfortunately, that is precisely when every other company is also looking for them, causing a quantum freeze in the entire supplier market.

Cybersecurity vulnerabilities

In stable times, cybersecurity becomes an afterthought. The focus shifts to efficiency, not defense, while the threats keep evolving. Your security system exists in a quantum superposition. It is both effective and compromised at the same time. You only discover its true state when an attack collapses this superposition, revealing a breach that was there all along.

Legacy systems are a special kind of risk. Their architecture exists in a superposition between adequate and dysfunctional. The systems function adequately until a business suddenly undergoes a rapid shift to remote work or adopts a new business model, necessitating an evaluation. Suddenly, the system's limitations collapse into reality, constraining the entire organization's ability to adapt.

Capital structure optimization risks

Low interest rates create a dangerous illusion about debt. A leverage ratio that looks conservative with rates at 3% becomes aggressive and crushing when rates change without warning to 5% during refinancing. Your capital structure isn't fixed. Your capital structure exists in a state that ranges from conservative to reckless. The interest-rate environment serves as the measurement that determines which state it collapses into, often resulting in devastating cost implications.

Debt covenants also exist in a superposition. During stable earnings, compliance seems automatic, and covenant margins look comfortable. But your true debt capacity is uncertain. A quantum drop in earnings acts as a measurement, testing those covenants and potentially collapsing your financial flexibility into a state of constraint or even default.

These hidden risks show how stable markets can unravel quickly. In the next section, we'll explore examples in the context of specific industries and regulations and see how they contribute to the creation of the illusion of wealth.

Industry-specific applications

The principles of quantum finance don't apply uniformly. They manifest in unique, fascinating ways across different industries, each with its own patterns of superposition and collapse. While industries obviously don't follow the laws of quantum physics, their financial uncertainties behave in strikingly quantum ways. This new viewpoint has the potential to significantly alter the process of making decisions.

Let's explore a few key sectors.

Manufacturing

Quantum effects are visible in a variety of manufacturing processes across the factory floor.

- **Inventory Valuation**: A pile of raw materials at Quantaland Manufacturing exists in a superposition of values: its historical cost, its replacement cost, and its net realizable value. Only the act of closing the books and choosing a method collapses this superposition into the single number on the balance sheet.
- **Capacity Utilization**: Is a production line efficient or idle? It exists in both states until the production schedule is set. The scheduling decision is the quantum measurement that forces the line into one reality or the other.
- **Supply-Chain Finance**: Payment terms are financially entangled with supplier health. A delay from one key supplier affects them and then ripples through the entire chain. During a cash crunch, this entanglement simultaneously collapses all flexible terms into strict, immediate demands.

Technology

The tech industry is built on quantum uncertainty.

- **Intellectual Property:** A patent's value is a classic superposition. It is both the sunk cost of its development and the potential future licensing revenue. This superposition only collapses into a definite value upon a sale or an impairment test.
- **R&D Spending:** Is it an expense or an asset? For most of its life, an R&D project is in a quantum state of both. Tech companies hold it in this gray area until a product launch or failure acts as the measurement event, collapsing it definitively into one category or the other.
- **Platform Value:** The worth of a digital platform seldom grows smoothly. It exists in a low-value state until user numbers hit a critical mass. Then,

it quantum-leaps to a new, much higher valuation level. The transition is sudden and discontinuous, not gradual.

Financial services

Finance is the quantum realm.

- **Credit Risk:** A loan on Quantaland Bank's books isn't just good or bad. It exists in a superposition of potential outcomes: full repayment, partial loss, or total default. This superposition of probabilities only collapses into a known loss when the borrower misses a payment.
- **Regulatory Capital:** A bank's capital position is constantly fluctuating. It exists in a state of both compliance and non-compliance until a regulator's stress test acts as the ultimate measurement. This forces the bank's capital reality into a single, definitive state.
- **Liquidity:** A company's capital is a quantum state. It seems sufficient until a crisis occurs. A liquidity crisis is the measurement event that collapses this superposition, instantly revealing whether the entity has enough cash to survive or not.

Healthcare

In healthcare, life-changing decisions are quantum events.

- **Drug Development**: Like R&D, a drug candidate is the ultimate quantum asset. For years, it has been neither a success nor a failure but a probability curve based on clinical data. The regulatory approval decision is the quantum measurement that collapses these years of long superposition into a multibillion-dollar success or a total write-off.
- **Company Valuation**: A biotech firm's value is binary. It is in a pre-approval state or a post-approval state. The regulatory decision causes a catapult between these two states, with no intermediate steps. The company's worth changes overnight.
- **Insurance Reimbursement**: Will a specific treatment be covered? Until the insurer rules, it exists in a superposition of yes and no. The coverage decision is the collapse that determines the financial reality for both the patient and the hospital.

Energy

The energy sector is volatile by nature, a perfect quantum system.

- **Reserve Valuation**: An oil reserve isn't a fixed asset. Its value is a superposition of possibilities based on future prices and costs. It could represent either a goldmine or a stranded cost. Only a final investment decision or a sale collapses this uncertainty into a firm number.

- **Commodity Prices:** Energy prices don't move smoothly. They exist in a "probability field" until a minor event, a storm, or a political quote triggers a sudden move to a new price level. Trading is the act of observing and collapsing this field into a definite price.
- **Environmental Liabilities:** The cost of cleaning up a site isn't known. It exists as a range of possibilities based on future regulations. An environmental audit is the measurement that collapses these possibilities into a specific, dreaded liability on the balance sheet.

Retail

Retail is a daily exercise in quantum collapse.

- **Inventory Value:** A dress on the rack at Quantaland Fashion is both full price and clearance price. Its value is in superposition until the customer's decision to buy, or the end of the season, collapses it into a final number.
- **Loyalty Programs:** Customer points are a quantum liability. They are both a future cost and a marketing asset. The redemption rate is the measurement that collapses this superposition, deciding if the program is a profit driver or a money dump.
- **Store Leases:** Is a store location profitable? Up until the decision to renew the lease, the store's profitability fluctuates between profitable and unprofitable states. The decision to renew or close is the quantum event that collapses this potential into a concrete financial reality.

Real estate

Real estate value is never settled until the deal is done.

- **Property Valuation:** A building's worth is a quantum scenario. It has multiple values based on income, comparables, or replacement cost. Only a sale transaction can collapse all these possibilities into one agreed-upon price.
- **Development Costs:** The final cost of a development project is unknown. It exists as a probability distribution over many scenarios. Only the act of building, with all its surprises, collapses this distribution into a final, often shocking, number.
- **Market Cycles:** Real estate markets don't turn gradually. Real estate markets undergo a sudden transition from a seller's market to a buyer's market. Transaction volume is the key indicator, the measurement that confirms the market's phase has collapsed into a new state.

Accounting, regulatory, and quantum reality

Our entire system of accounting and regulation is built on a classical foundation, the idea that a company's financial position is an objective fact waiting to be measured and reported. Quantum finance challenges these premises. It argues that uncertainty isn't a problem to be solved but a fundamental feature to be managed. Financial outcomes exist within a vast array of possibilities. Only a specific event – a sale, a default, or a court ruling – can force them to collapse into a single reported number.

Failure to grasp this will leave us perplexed by unexpected corporate collapses. The financial reports didn't lie. They simply reported the reality that existed at the reporting date, completely overlooking other potential realities that were equally possible and later materialized. The core failure is that our systems treat wealth as permanent when it is, in fact, profoundly transitory.

IFRS/GAAP implications

Accounting standards such as IFRS and GAAP are a brave attempt to tame uncertainty. They use probability-weighted estimates and three levels of fair value to try and pin down a "true value." This approach assumes value is a hidden fact to be discovered. The quantum view is different, and it sees value as something that drifts across a range of outcomes. The act of measurement itself helps create the final value.

Fair-value accounting gets closest to acknowledging this. It allows for a range of inputs and requires sensitivity disclosures. But eventually, the accountant must pick one number for the balance sheet. This number is like a quantum wave-function collapse. It is just one of many possible states that happened to become real at the reporting date. It is not the absolute truth, merely a truth that holds for the time.

Look at revenue recognition. The rules have evolved to handle variable considerations, but they still assume a contract can define revenue. Quantum finance says revenue isn't real until the customer acts by paying, delaying, or even canceling. The customer's action represents the collapse event.

This is why you must never look at a financial statement from Quantaland, or anywhere else, as sacrosanct. You are looking at a snapshot of a single reality that emerged from a sea of possibilities at a specific moment. By the time you read the report, that moment has passed, and a new set of possibilities has already emerged. Unfortunately, you are not aware of it.

Regulatory requirements

Regulators face an impossible task. They formulate regulations grounded in the traditional notion of identifying, measuring, and managing risk.

Banking capital rules and regulations assume Quantaland Bank's risks are known and quantifiable. But its true risks, such as a synchronized market crash or correlated defaults, exist in the superposition of hidden states. They only collapse into devastating reality during a crisis – rendering capital ratios instantly inadequate.

Insurance solvency models use probabilities to estimate future claims, which acknowledges some uncertainty. But they still imply that the final cost can be known. Any seasoned insurer in Quantaland will tell you the final bill is always a surprise. Liabilities remain in a fluid, probabilistic state until the very last claim is settled.

Risk disclosures for public companies end up being a list of risks in their annual reports, posing as discrete, understandable items. This is a classical illusion. Significant business risks are a tangled web of interconnected possibilities. They remain in complex superposition until a market event or a decision forces one specific risk pathway to become real.

Tax considerations

Taxation is a world of enforced certainty, but it's built on quantum uncertainty. Please take into account the transfer pricing rules. These rules pretend that a single "arm's-length price" exists for transactions between related companies. In reality, a range of prices could be justified. This is a tax superposition. The price only collapses into a single, "correct" number when a tax authority examines it. It is common for the same transaction from Quantaland International to trigger different tax bills in various countries, as each tax authority interprets the superposition differently.

Companies book provisions for uncertain tax positions, acknowledging a range of possible outcomes. But deep down, the system believes there is one right answer. Quantum thinking says true tax obligation arises only when an audit or a court case measures it. This measurement creates liability for tax.

This shows how our entire concept of wealth assumes that a single, objective reality exists. When the real world, which is quantum, exposes the flaw in this assumption, we react with surprise. We blame the event, not our invalid starting point.

Audit standards

After a corporate collapse, the first question is always, "Where were the auditors?" The answer often lies in quantum uncertainty. Auditing is a classical process in a quantum world.

Auditors look for risks in neat categories based on past data. However, quantum risks are evolving and consist of interconnected possibilities.

A single audit risk represents a superposition of potential future outcomes. The audit process is not designed to see this cloud of possibility. An audit only examines the reality that existed at a point.

Auditors set numerical thresholds to decide what is important. But materiality is not a fixed number. It exists in the superposition of perspectives. What is immaterial to one investor could be critical to another. The auditor's measure is just one collapse of this superposition.

An auditor's opinion on whether a company will survive is one of the ultimate quantum judgments. The business exists in multiple possible futures. Will it secure funding? Will it lose a key customer? The auditor uses models to make a call, but the true outcome remains in superposition until a specific event, like a deal falling through, collapses it into survival or failure.

We have seen how the quantum world creates the illusion of permanent wealth across industries. Now, let's turn to the headlines, to events such as the GameStop surge and crypto crashes, to see these quantum principles playing out in real time in the most dramatic ways possible.

Meanwhile, far from Quantaland

The GameStop Short Squeeze (2021)

The Quantum State: GameStop's stock embodied two valid views, one of a failing retailer and the other of transformative opportunity. This range of possibilities, like a quantum system, collapsed into wild price swings as market actions unfolded. The market price fluctuated wildly as these competing views came into prominence.

The Story: The GameStop saga of early 2021 represents one of the most dramatic demonstrations of wealth transience in modern financial history. GameStop, a brick-and-mortar video game retailer facing declining sales and pandemic-related store closures, became a favorite target of short sellers. By January 2021, short sellers had sold more shares than GameStop's total outstanding shares, a situation that would later prove crucial. You know the short sellers – those who sell a share expecting the value to go down and who would make a profit by buying those shares at a lower price and honoring their commitment. In contrast, if the price increases, they incur a loss. Sometimes they would buy early to limit their losses.

The story took shape on the Reddit forum, where retail investors led by Keith Gill (known online as "Roaring Kitty") argued that GameStop was significantly undervalued. Gill had been publicly sharing his bullish thesis since 2019, highlighting the company's strong cash position, potential for digital transformation, and the upcoming console cycle as catalysts for revival. This view gained further momentum in September 2020, when Ryan Cohen, co-founder of the successful e-commerce company Chewy, disclosed a significant stake in GameStop. He joined the board in January 2021. Cohen's involvement was interpreted as a possibility for a digital transformation for the struggling retailer.

Millions of retail investors, coordinated through social media, began purchasing GameStop shares and call options in January 2021, reaching a critical mass. Surely you recall that a call option is a derivative instrument that gives you the right to buy the share at a decided price on or before a decided date. Of course, you may decide not to buy. This buying pressure, combined with the short interest, triggered a massive squeeze. Hedge funds that had bet against GameStop were forced to buy shares to cover their positions, further driving up the price in a self-reinforcing cycle.

The results were spectacular. GameStop's stock price rose from approximately \$17 at the beginning of January to a peak of \$483 on January 28; you can do the growth arithmetic. At its height, GameStop reached a market capitalization of over \$24 billion, creating and destroying billions in wealth in a matter of days.

The situation changed dramatically on January 28, when several brokerages, most notably Robinhood, restricted buying in GameStop and other heavily shorted stocks, citing clearinghouse margin requirements. This controversial decision effectively halted the momentum and led to a rapid decline in GameStop's stock price. It fell to around \$40 by mid-February. If you find some other price, most likely it is after the 4:1 stock split.

The Quantum Finance Lesson: The GameStop saga vividly illustrates how wealth can exist in a state of superposition – similar to quantum phenomena. In this scenario, GameStop's value was not fixed but instead fluctuated between radically different valuations, each supported by distinct narratives and perspectives. Like quantum particles that do not assume a definite state until observed, the "true value" of GameStop remained indeterminate until market actions forced a resolution. As these various viewpoints gained prominence, the stock's value swung dramatically, showing the uncertainty inherent in financial markets.

The coordinated buying efforts of retail investors created a powerful feedback loop with institutional short sellers, magnifying the volatility of the stock price. This interplay can be likened to quantum entanglement, where the actions of one participant instantly affect the state of the other. In the case of GameStop, the mutual influence between retail investors and institutional players led to rapid shifts in collective market perception, causing wealth to transition abruptly between different states. This episode highlighted a fundamental aspect of quantum finance: collective observation and action can swiftly shift wealth, a reality not always accounted for by traditional financial models.

Practical Application: Investors should recognize that market valuations represent probability distributions rather than fixed realities. When extreme positioning occurs, the potential for quantum-like transitions increases dramatically. This suggests maintaining flexibility in investment strategy and acknowledging that both bulls and bears can be simultaneously right and wrong, reflecting a quantum-like superposition.

The Flash Crash

The Quantum State: Markets balanced normal trading and chaos, like a quantum system with multiple possible states, until algorithmic selling forced a sudden collapse. Prices simultaneously reflected both rational valuations and those generated by algorithmic feedback loops.

The Story: On May 6, 2010, U.S. financial markets experienced one of the most dramatic and puzzling events in trading history. In just 36 minutes, between 2:32 p.m. and 3:08 p.m. ET, the Dow Jones Industrial Average plunged about 1,000 points (9%). Then, within minutes, the Dow Jones Industrial Average mysteriously recovered most of its losses. This "Flash Crash" wiped out nearly $1 trillion in market value momentarily before the mysterious rebound.

The day had begun with markets already under pressure due to concerns about the Greek debt crisis. Around 2:32 p.m., an unusually large sell order in futures contracts was placed by Waddell & Reed, a mutual fund company. The order was for 75,000 E-Mini S&P 500 futures contracts valued at approximately $4.1 billion. This acted as a key trigger for the crash, amplified by high-frequency trading and market stress.

Apart from the size, its execution was another extraordinary aspect of this trade. The trading algorithm was programmed to sell based on trading volume without regard to price or time. When the algorithm began selling futures contracts rapidly, high-frequency trading firms initially stepped in to buy. However, they quickly reversed course and began selling as volatility increased.

If you ever wanted to know what nonsense is, here it was. The selling spree drove more sales. Liquidity evaporated as market makers either widened spreads or withdrew entirely. In the vacuum created by this liquidity withdrawal, absurd outcomes materialized. You may not believe that shares of Accenture briefly traded for one cent due to placeholder bids, while other stocks saw extreme swings as liquidity vanished.

The recovery was as mysterious as the crash. Around 2:45 p.m., trading on the Chicago Mercantile Exchange paused for five seconds due to circuit breakers being triggered. This brief pause allowed enough liquidity to return to the system, and a remarkable recovery began. By 3:08 p.m., most indices had regained their pre-crash levels, though the market closed lower for the day. Numerous simultaneous actions played a significant role in this event; had you discussed this phenomenon earlier, you would have faced ridicule.

The Quantum Finance Lesson: Financial markets can behave in ways reminiscent of quantum systems, where order and chaos coexist and abrupt transitions happen with little warning. The Flash Crash vividly exemplifies this phenomenon, as the market rapidly shifted to an unexpected state and prices briefly reached levels considered impossible by traditional financial models. This event reminds us that conventional approaches often underestimate the likelihood of such quantum-like transitions.

During the Flash Crash, market liquidity itself displayed quantum behavior – existing more as a probability function than a certainty. Under ordinary conditions, markets seem continuous and liquid, yet beneath this stability lies a discrete, quantum nature, where liquidity can vanish in an instant and prices can jump unpredictably.

Practical Application: You should incorporate quantum thinking into risk management and will be better off recognizing that market continuity is an illusion. This illusion is susceptible to sudden collapse. To manage such volatility, incorporate quantum thinking into risk management. Use stress tests to simulate sudden liquidity drops, like those in the Flash Crash. You may set position limits to cap exposure to algorithmic swings and diversify portfolios to buffer against rapid market shifts. You need to recognize that volatility is not an exception but a feature of complex adaptive systems.

Cryptocurrency volatility

The Quantum State: Cryptocurrencies balance multiple identities such as speculative assets, technological innovations, and potential currencies, with values swinging wildly as these narratives shift.

The Story: Bitcoin, the world's first decentralized cryptocurrency, has exhibited some of the most dramatic price volatility in financial history. The volatility demonstrated the transitory nature of digital wealth. Created in 2009 by the pseudonymous Satoshi Nakamoto in the aftermath of the global financial crisis, Bitcoin was designed as a peer-to-peer electronic cash system that operates without central authority.

Bitcoin's price swings feel like a quantum puzzle, balancing multiple values. From 2009 to 2011, it traded below $1, barely noticeable. By February 2011, it hit parity with the U.S. dollar. The real drama came later: Bitcoin surged to nearly $20,000 by December 2017, only to crash to around $3,200 by December 2018. You must have calculated the 84% drop. Three years later, it soared to an all-time high of nearly $69,000 in November 2021 before plummeting to about $16,000 in November 2022 after the FTX collapse. The volatility didn't stop there – Bitcoin reached a new peak of over $103,000 in December 2024, showing how its value can swing wildly, like a quantum system balancing multiple states until market events pin it down.

The volatility extends beyond Bitcoin to the broader cryptocurrency ecosystem. Ethereum, the second-largest cryptocurrency, surged from around $10 in early 2017 to over $4,800 in November 2021 and then dropped to about $900 in June 2022. It later rebounded, trading around $2,400–$3,000 in mid-2025. The rise of "memecoins" has been even more dramatic. Dogecoin was originally created as a joke. Its value surged from about $0.0047 on January 1, 2021, to a peak of $0.69–$0.74 in May 2021. After this gain of roughly 15,500–16,700%, it plunged over 90% to around $0.049 in June 2022. In May 2022, Luna's market cap crashed from $40 billion to nearly zero in a week as its stablecoin UST lost its peg, devastating investors.

Perhaps most telling was the rise and fall of Luna and its associated stablecoin UST in May 2022. In just one week, Luna's value collapsed from over $40 billion to virtually zero as its algorithmic stability mechanism failed, wiping out the savings of numerous investors. People went to sleep as millionaires and woke up broke.

The Quantum Finance Lesson: Cryptocurrencies provide a vivid demonstration that digital assets can exist in a kind of superposition of value. Their prices do not reflect a single, fixed reality but rather represent a shifting probability distribution of potential future outcomes. Instead of being anchored to inherent worth, cryptocurrency values are shaped by collective beliefs about what they might become. Like Schrödinger's cat, cryptocurrencies inhabit a space where they are both groundbreaking innovations and speculative bubbles until market forces, like observation in quantum mechanics, resolve these possibilities into one reality.

Significant market events – regulatory developments, advances in technology, or shifts in sentiment – act as observations that cause a cryptocurrency's value to settle into a definite state. In practice, the journey to determine genuine value in cryptocurrency is ongoing. Progress in adoption comes in cycles: speculation, collapse, and technological advancement gradually reveal each asset's lasting utility.

Practical Application: Investors approaching cryptocurrencies should adopt quantum thinking. You must recognize that traditional valuation methods are insufficient for assets whose value depends on multiple potential future states. This means approaching crypto investments with position sizes that acknowledge the possibility of both zero value and exponential growth. This calls for a probabilistic rather than deterministic approach.

Bitcoin even today remains in a transitional state. It is simultaneously a speculative asset today and a potential store of value tomorrow. The real worth of Bitcoin will emerge when the dust settles over the transition.

Looking ahead

We've seen how wealth exists in quantum superposition. It is simultaneously valuable and worthless, stable and volatile, until the moment of measurement forces it into a single reality. However, this poses a profound question at the core of financial analysis: what happens when the process of measuring wealth itself alters its value? The Flash Crash showed us how algorithmic measurement created chaos. GameStop demonstrated how collective observation shifted reality. In our next chapter, we'll explore this quantum measurement effect in depth. We will discover how financial reports don't just record reality; they actively shape it. You'll learn why the Brexit referendum's polling data influenced its outcome and why your company's earnings guidance doesn't predict the future but actually helps create it. Get ready to challenge all your previous beliefs about how financial measurement relates to financial reality.

Chapter 4

The measurement effect: how observations impact decisions

"*Shadows cast by moonlight change how the wolf looks*," goes an old Quantaland proverb.

A bullish report by analyst Stephen had set the stocks of a retailer soaring on a path of rise until a sudden sell-off erased the gains. Investors had overreacted to his report, Stephen thought. His observation didn't just reveal; it actively reshaped reality. This chapter shows how financial observations create quantum-like feedback effects. It will explore why earnings, ratings, and measurements shape performance through self-reinforcing cycles. You will also examine cases like Brexit and vaccine announcements and understand why markets move dramatically beyond fundamentals. This understanding will help you navigate a world where observation is a force of change.

The active observer: when measurement creates reality

Classical finance operates on a simple assumption that a company has a "true value," and our analysis simply uncovers it. Quantum finance challenges this notion. It argues that observing financial data, a stock price, or an earnings report does much beyond passively revealing reality. The act of observation actively changes it. Investors, managers, and entire markets react to what they see, making decisions that alter the very outcome they are measuring. This isn't just irrational behavior, as some in behavioral finance might argue. It is a fundamental feature of the system. Measurement itself triggers a feedback loop of action that pushes markets in unexpected and often unpredictable directions.

The reflexive loop: how markets shape their own reality

George Soros's theory of reflexivity is the perfect bridge to quantum finance. It argues that markets are not passive mirrors reflecting economic reality. They are active participants, shaped by how people perceive data and then

DOI: 10.4324/9781003619635-4

act on those perceptions. When you measure a financial outcome, like a company's earnings or a property's price, you don't just receive a number. You trigger a reaction that changes the entire game.

Look at Quantaland's housing boom. Rising prices weren't just a signal; they were a catalyst. They convinced investors that the rise would continue, so they poured in more money, driving prices higher still. Seeing these ever-higher valuations, banks relaxed their lending standards, further exacerbating the situation. This self-reinforcing loop continued until the central bank raised rates. Then the loop reversed. Falling prices shattered confidence, banks tightened lending, and values spiraled down. The act of measuring and reporting property values didn't just capture market conditions; it actively drove them in both directions.

This reflexive cycle is universal. Currency crises are a classic example. If investors fear that the currency will fall and start selling, their sales trigger the fall, validating their fear and triggering more sales. The same principle applies to a booming sector like AI in Quantaland. Investors' belief in the potential of the sector drives investment. That investment fuels growth and innovation, which justifies the initial belief and attracts more capital. The measurement of success, in the form of rising valuations, creates the very success it is measuring.

Financial reporting as market observation

Quarterly earnings reports serve as more than just scorecards; they are transformative tools. In Quantaland, when a company like Quantaland Software releases earnings, it's not just reporting numbers. It's setting off a chain reaction. Investors bid up or sell off the stock, and management tweaks their playbook based on that response.

Beat expectations, and management might double down on growth – say, by boosting R&D or expanding into new markets. Miss the mark, and they could slash long-term projects, like introducing a new product line, to chase short-term gains for the next quarter. Revenue recognition choices, like deferring income to smooth earnings, can amplify these effects. Such acts would not only shape market perceptions but also end up influencing actual cash flows and strategy.

These reactions function as feedback that influences future performance, creating a self-reinforcing cycle. Strong results can lead to continued growth, while poor results can start a downward spiral. Management behaviors respond directly to the impact of reported outcomes.

The effect also ripples out. Strong earnings boost employee morale, improve supplier relationships, and help attract new customers and talent. Meanwhile, annual reports magnify the impact. Annual reports extend beyond mere numbers, compelling management to elucidate their vision and strategy, ultimately shaping the implementation of that strategy in practice.

Audit and assurance reflexivity

Audits don't just check the financial statements; they end up reshaping the business. At Quantaland Systems, the mere announcement of an audit acts like a quantum measurement, bringing processes into focus. Months before auditors arrive, teams tighten controls, review revenue recognition, and double-check inventory counts. You can clearly see that the audit is not just compliance; it brings a shift in how the company operates. Let's look deeper into this cycle.

When auditors zero in on, say, revenue recognition, management ramps up oversight of sales contracts. A clean audit opinion can unlock cheaper loans or boost investor confidence, fueling growth. However, what would happen if the audit opinion was qualified? It's like a quantum collapse. Lenders tighten terms, suppliers get jittery, and suddenly, the company's future looks uncertain. Audits don't just report reality; they help create it.

On the other hand, a qualified opinion may trigger a chain of challenges. It can raise red flags with lenders, suppliers, and partners. This can even trigger the very issues the audit opinion warned about. For example, a qualification on inventory control issues may prompt banks to tighten their funding, leading to a major crisis in inventory management. In this way, audits become more than assessments; they help shape the company's future path.

Credit-rating reflexivity

Credit ratings, beyond reflecting a company's financial health, play a role in shaping it. When Quantaland Corporation secured an investment-grade rating, it gained access to a broader pool of institutional investors, which lowered its borrowing costs. This, in turn, improved their cash flow and strengthened their financial standing, which is most likely to help them maintain, if not improve on, the current rating.

Virtuous cycles like this turn vicious in the case of a downgrade. If the rating slips from BBB+ to BBB, some investors are likely to withdraw, borrowing will be expensive, and covenant breaches will become more probable. The resulting strain on cash flow can lead to further downgrades, triggering a downward journey.

Companies often adjust their strategies to align with the key metrics of rating agencies. They might hold off on acquisitions or accelerate asset sales solely to keep ratios like debt-to-equity or cash-flow coverage in line with rating preferences. Rating reflexivity rules over operational prudence.

The timing of rating changes adds another layer. Rating agencies often delay action until obvious signs of trouble appear. This means that downgrades happen only when companies are already vulnerable. The ratings lose their lead indicator quality, and the consequent shock value enhances financial stress.

Management guidance and forward-looking statements

Management guidance is a clear example of how financial observation influences real-world outcomes. When executives issue earnings forecasts or strategic projections, they do more than share expectations. These forecasts shape internal decisions and market behavior in ways that impact the results themselves.

For instance, when Quantaland Software projects Q$50 million in quarterly revenue, that figure becomes a target for the entire organization. To align with the revenue recognition timelines, sales teams adjust their tactics, realign marketing budgets, and reschedule product launches. Management guidance becomes a driver of operational choices ahead of the actual earnings announcement.

Positive forward-looking statements also influence external perceptions. Customers and suppliers respond positively to strong projections, often offering favorable contract terms or partnerships.

As a company continues its dream run, repeatedly meeting or exceeding its guidance, investor expectations rise. This creates pressure on management to keep delivering stronger projections, leading to increasingly optimistic targets. This comes with a risk. To meet short-term goals, leadership may postpone long-term initiatives. Research and development timelines, strategic partnerships, and investments may all be influenced more by the guidance calendar than by long-term value.

Regulatory compliance reflexivity

Regulatory oversight does not merely monitor corporate behavior; it changes the behavior. Regulations create measurement effects that influence how businesses operate, often before any formal findings are issued.

At Quantaland Bank, increased scrutiny of lending practices leads to proactive changes. The announcement of a regulatory review triggers change. The loan officers tighten lending criteria, risk models are reviewed, and paperwork gets more detailed. These are all done in anticipation of what regulators might expect, not necessarily what the regulations demand.

The cost of compliance also drives strategic decisions. Often, regulations make certain activities pricier, primarily through enhanced compliance requirements. These companies may restructure their operations to reduce the regulatory load. Strict requirements may motivate a company to be more efficient in compliance than in customer experience or operational quality.

Often, firms implement short-term enhancements during review periods, only to relax them once the period concludes. This creates cycles of artificial compliance that are periods of temporary "compliance reality," rather than lasting reform.

Regulatory reporting requirements impact how resources are allocated. When certain metrics must be reported, they will be prioritized even if they don't align with core business goals. The burden of measurement can lead to this behavior. In areas such as environmental compliance, firms may favor simpler, easier-to-measure solutions over more effective but harder-to-document alternatives.

Feedback loops: how investor reactions amplify market movements

Financial markets are not passive absorbers. They are active engines. When investors react to a piece of data, a stock price, or an earnings report, their actions don't just respond to the market; they become new market data themselves. This creates a feedback loop where observations fuel reactions, amplifying tiny ripples into massive waves.

This explains the enormous puzzle of why markets consistently overshoot either way. Traditional finance believes prices adjust efficiently to the latest information. Quantum finance sees it differently. It argues that the act of measuring the market changes it. Investor reactions to a price create self-reinforcing cycles that drive valuations far beyond anything the original news could justify.

The amplification engine: how reactions become market data

Feedback loops occur when market participants respond to information in ways that change the underlying conditions that create it. Unlike simple cause-and-effect relationships, these loops create circular dynamics where effects become causes, and causes become effects. In essence, the market's reaction to a piece of information becomes the latest information that influences subsequent reactions. This creates self-reinforcing cycles that can persist until external forces break the pattern or the cycle reaches unsustainable extremes.

Earnings management feedback cycles

Traditional finance sees an earnings report as a report card. Quantum finance views it as a crucial tool. At Quantaland Electronics, beating quarterly targets boosts the stock price. This rewards management and tempts them to do whatever it takes to hit the next number, cutting costs indiscriminately, pulling sales forward, or selling assets.

This works in the short term. Options vest, and everyone feels successful. But with each success, expectations rise. The pressure to perform forces management to make devastating long-term choices, delaying R&D and slashing marketing to protect the quarterly profit. Each earnings report is like a quantum measurement, collapsing the company's range of future strategic options into a narrow path focused only on the next quarter.

This cycle continues until it breaks. When the company finally misses its target, the very feedback loop that created success massively amplifies the market's negative reaction. This triggers a painful collapse, resulting in widespread suffering.

Capital-allocation feedback effects

A company's stock price isn't just an output; it's a key input for strategy. At Quantaland Industrial, a rising stock price validates management's strategy, boosting their confidence. A high valuation makes acquisitions cheaper (using stock as currency) and makes banks, suppliers, and customers more eager to collaborate with them. This creates a virtuous cycle. Robust performance fuels investment in R&D and expansion, which fuels further powerful performance.

But this loop has a dark twin. A falling stock price makes management cautious, prompting them to delay investments and scale back growth plans. However, this cautiousness weakens the company's competitive position, leading to poorer performance and further declines in the stock price. The falling price makes it harder to attract talent and retain stakeholder confidence, accelerating the decline. The feedback loop that built the company up can now just as efficiently tear it down.

The media and analyst echo chamber

Financial media and analysts don't just report on companies; they reshape them. Positive coverage attracts investors, boosts morale, and helps recruitment. In response, management starts shifting resources and strategy toward the very areas that are getting praised. The analysts' favorite projects get funded.

But when coverage turns negative, management often panics. Their public response can amplify the concerns, forcing them into abrupt restructuring or strategy changes they hadn't planned. This shows that analyst calls and media reports aren't just background noise; they are quantum measurement events that collapse a range of strategic options into a single, reactive path.

The questions analysts ask during earnings calls are particularly powerful. They signal what the market cares about. If every analyst asks Quantaland Solutions about its AI strategy, you can bet the management will soon have a bigger AI strategy to talk about. The rise of social media has supercharged this effect. Companies now monitor sentiment in real time and calibrate their story, creating a frantic, never-ending feedback loop between perception and action.

Stakeholder engagement loops

Companies are in constant dialogue with their stakeholders, and this conversation directly shapes what they do. At Quantaland Retail, investor relations calls aren't just for updates. They are a convenient channel for investors to tell management what to focus on, directly influencing which business segments get resources.

Shareholder activism turbocharges this effect. To avoid a painful proxy fight, management will often preemptively adopt an activist's demands for cost cuts, spin-offs, or governance changes. It's crucial to distinguish between actions that genuinely improve the business and those taken simply to placate loud shareholders.

This feedback loop extends to customers and employees. In addition to enhancing product quality, customer feedback guides the selection of products for production. Employee surveys can trigger shifts in management style and compensation. Recently, ESG has become the ultimate stakeholder feedback loop. Companies are increasingly shaping their entire business model around these external metrics, sometimes at the expense of traditional financial performance. The stakeholders have ceased to remain just observers. They are the new architects.

Earnings report ripple effect

The relationship between earnings announcements and market reactions creates some of the most visible measurement effects in finance. These reactions don't just reflect cold assessments of historical performance. They actively shape future business decisions and market dynamics in real time, proving that in finance, observation is never passive.

Earnings surprises

Earnings surprises are a perfect example of the quantum measurement effect. Markets don't just react to the news; they react to the difference between the news and the expectation. This often causes massive overreactions that have little to do with the actual monetary impact. This overreaction then forces management to change their behavior, often creating a self-reinforcing cycle, for better and for worse too.

Quarterly reporting cycle effects

The relentless tick of the quarterly clock often forces businesses to operate against their natural rhythm. Quantaland Aerospace breaks down the intricate, year-long airplane-building process into 90-day reporting periods. Delivery schedules are jerked around to make a quarter look satisfactory, sacrificing long-term efficiency for a short-term accounting win.

This quarterly obsession distorts everything. Sales teams give away the store to close a deal by quarter's end. Factories overproduce to flatter inventory numbers. Vital research is delayed or rushed to align with an arbitrary calendar. This obsession, besides painting a distorted picture for the market, nurtures a distorted strategy inside the company.

The constant measurement also amplifies noise. Missing one quarter can trigger a panic, while beating expectations can create dangerous arrogance. Worst of all is this focus on a three-month sliver of time that means extraordinarily little in the life of a significant business.

Revenue recognition timing impacts

Revenue recognition rules are a powerful example of the measurement effect. The timing of when revenue is booked often moves markets more than the underlying business performance. This forces companies to structure their entire operations around accounting treatment, often preferring it over customer needs.

At Quantaland Software, teams scramble at quarter's end to renew customer contracts. They offer discounts for early payments or change service schedules, all to pull revenue into the current quarter. If the customer gets a better deal, it's a joyful accident. The primary goal is to hit the number.

This creates a vicious feedback loop. Instead of governing reporting, the accounting practice begins to design business models. Customers soon learn this game, delaying purchases until the quarter-end fire sale begins. This creates artificial demand cycles that crush profit margins. The accounting department begins to exert significant influence on the business.

Non-accounting metrics and adjusted earnings

When companies can't win under the standard accounting rules, they simply invent new ones. "Adjusted" earnings and non-accounting metrics are a direct response to measurement pressure. They don't just add information; they create an alternative reality designed to influence perception and justify behavior.

Quantaland Technology's "adjusted" earnings famously exclude the costs of acquisitions, stock-based compensation, and restructuring. This isn't clarification; it's propaganda. It encourages reckless acquisitions and excessive pay because the metrics are engineered to ignore the cost. The framework is built to achieve the desired market reaction, not to reflect business fundamentals.

This has become so pervasive that accounting standard setters are fighting back. The new IFRS 18 standard on "Management Performance Measures" is a direct regulatory response to this proliferation of fantasy metrics. However, this creates a new measurement effect because companies will now modify their strategies to align with these new disclosure norms.

The guidance trap

Management guidance is a trap. It starts as a forecast but quickly becomes a commitment. The original intent, to communicate expectations, is inverted. Soon, business decisions are made primarily to meet the guidance, not to serve the strategy.

The accuracy of a company's guidance becomes crucial to its market credibility. This forces management to pretend they can predict the unpredictable. Companies that play this game well are rewarded with higher valuations, which only encourages more companies to play.

The pressure isn't uniform. A tech company like Quantaland Software is expected to provide ambitious growth guidance, pushing them toward risky bets. A stable utility like Quantaland Power is expected to provide boring, reliable guidance, pushing them toward extreme conservatism. In both cases, the guidance becomes a central operating principle, dictating supplier terms, customer deals, and employee bonuses. Like a quantum measurement, the act of providing guidance collapses the company's future possibilities down to a single, restrictive path.

Industry and sector rotation effects

An earnings surprise, in addition to affecting one company, infects the entire sector. This is the measurement effect in its most contagious form. The market uses one company's results as a proxy to measure all its peers, regardless of their individual fundamentals.

The ripple effects spread throughout the business ecosystem. A strong report from Quantaland Automotive, together with boosting its stock, instantly props up its suppliers and punishes its competitors, all before they report their results. A bad report can trigger a sector-wide panic, hammering companies that have no direct connection to the original news. The market no longer analyzes companies. They start diagnosing them in clusters based on a single data point.

Global and macroeconomic implications

The measurement effect of earnings is so powerful that it can sway entire economies. The earnings report of a single multinational can move currency markets, as traders use it to measure the health of a country's trade flows. This is a stunning inversion, where a corporate result influences the value of the national currency.

This goes even further. Central banks watch aggregate corporate earnings as a key input for interest-rate decisions. Governments shape tax policy based on corporate profitability trends. This creates a massive feedback loop where corporate measurement effects directly influence public policy, which then loops back to shape future corporate behavior. The decision to invest in a country or abandon an industry can hinge on these feedback loops. The quarterly earnings report, a seemingly micro event, thus becomes a macroeconomic force.

When sentiment dictates strategy

Sentiment indicators are the mood scale of the market. These tools go beyond merely measuring collective investor psychology; they actively reshape it. These tools create powerful measurement effects that directly influence both market behavior and corporate strategy, proving that gauging sentiment is never a passive act.

Corporate sentiment measurement

Every word a CEO utters on an earnings call is now measured for sentiment. The tone can often move markets more than the numbers. Analysts run the audio through algorithms, dissecting language patterns for hidden meaning. This creates a feedback loop where the measured sentiment dictates the executives' next move.

If sentiment analysis detects unusual optimism from the CEO of Quantaland Energy, it will instantly trigger algorithmic trades. Management knows this. They consciously craft their messaging, choosing words designed to score well on sentiment indicators.

This effect extends to surveys. A CFO confidence survey not only measures sentiment but also influences it. CFOs who participate become more conscious of the mood and start adjusting their plans and predictions to align with, or counteract, the trend.

Even board appointments are sentiment signals. The market reads the news of the appointment of a new director or executive as a signal of strategic change, moving the stock price long before that person even steps into the office. The announcement itself is a measurement that collapses a range of possibilities into a new market reality.

Financial statement sentiment analysis

The annual report is now written for two audiences: humans and algorithms. Investment firms use deep learning to scour every word of these filings for sentiment cues. This has triggered a fundamental shift. Companies now craft their disclosures not for clarity, but to optimize their sentiment score with machines.

If Quantaland Corporation finds that phrases such as "economic headwinds" or "competitive pressures" trigger negative algorithmic scores, they will remove them from future reports. They will replace them with algorithm-friendly terms such as "growth opportunities" and "market leadership."

This creates a bizarre measurement effect where the substantive content of regulatory filings is shaped by how machines will feel about it. This isn't theoretical. Sentiment scores from these filings are now used in credit-rating models. The language in your annual report can literally change your cost of borrowing. We are no longer reporting mere financials; we are also reporting to please the machines that judge them.

Technology and digital sentiment tracking

Brands use social media sentiment as a real-time alarm system. Companies now monitor these channels constantly, and the feedback loops are instantaneous. This real-time measurement provides a raw, unfiltered insight that traditional research can seldom match.

Quantaland Consumer Products has a dedicated social media surveillance team. The moment they detect a negative trend, they pivot. Advertising campaigns are pulled, product positioning is tweaked, and spokespeople are dumped. Within hours of a scandal involving their celebrity endorser, the company removed the endorsement from the internet. Customer sentiment directly dictates immediate business decisions in this fast-paced feedback loop.

This extends to online reviews, which have become a powerful sentiment currency. We all choose hotels and restaurants based on stars and ratings. These scores have a direct impact on customer acquisition costs and lifetime value. Even measures of digital engagement, like time spent on an app, are now sentiment indicators that drive multimillion-dollar technology investment decisions. The digital mood of the crowd is now a core strategic input.

Credit and liquidity sentiment measures

Credit isn't just about numbers anymore; now it's also about mood. The swings in credit spreads are powerful sentiment indicators that directly dictate corporate financing strategy. Companies don't just time their debt issuances based on interest rates; they time them based on the market's emotional state toward credit.

The CFO of Quantaland Industries doesn't just look at the company's balance sheet. She obsessively monitors surveys on bank lending sentiment. A dip in lender optimism will cause them to draw down credit lines and hoard cash, not because their financials have changed, but because they fear the lending mood is about to shift. Financing decisions are driven by perception, not reality.

This extends to global strategy. Sentiment about a country's political stability, scraped from news analysis, will guide decisions on billion-dollar investments and government contracts. These sentiment indicators become a form of intelligence that shapes strategy at the highest level. Since everyone acts on these signals, sentiment becomes self-fulfilling. The fear of a credit crunch can itself cause one. The market's mood, on one hand, reflects reality. On the other hand, it creates it.

Meanwhile, far from Quantaland

Tesla's Bitcoin purchase and sale (2021–2022)

The Quantum State: Tesla's Bitcoin investment existed as both a corporate treasury strategy and a speculative position, with Musk's public statements serving as "measurements" that influenced the market perception of Bitcoin and its price.

The Story: In February 2021, Tesla announced it had purchased $1.5 billion worth of Bitcoin. This disclosure, made in a Securities and Exchange Commission (SEC) filing, represented approximately 8% of Tesla's cash reserves at the time. It was also one of the largest corporate investments in cryptocurrency.

The market reaction was immediate and dramatic. Bitcoin's price surged more than 20% on the news and reached the then-record high above $47,000. The investment was considered legitimizing Bitcoin as an institutional asset class.

However, Tesla's relationship with Bitcoin proved volatile. In March 2021, Musk announced that Tesla would accept Bitcoin as payment for its cars, contributing to Bitcoin going up to $58,000 by late March. However, in May 2021, Musk reversed course, stating concern over the rapidly increasing use of fossil fuels for Bitcoin mining and transactions.

This announcement triggered a 15% decline in Bitcoin's price within 24 hours. The price came down from around $57,000 to approximately $48,000 and erased hundreds of billions in market value across the cryptocurrency ecosystem. In June 2021, Musk declared that Tesla would begin to accept Bitcoin once mining transitioned to more sustainable energy sources. In July 2021, Tesla disclosed in its Q2 earnings report that it had sold approximately 10% of its Bitcoin holdings and had realized around $128 million in profit. Tesla presented this action as a test of Bitcoin's liquidity and promoted it as an alternative to holding cash on the balance sheet.

The story continued into 2022 when Tesla announced in its Q2 earnings report that it had sold 75% of its remaining Bitcoin holdings and had added $936 million in cash. The disclosure stated that the company had converted its Bitcoin into fiat currency, reducing its holdings from $1.26 billion to $218 million. Musk explained that this was done to maintain overall liquidity for the company, given COVID shutdowns in China, and they have not sold any of the Dogecoin investment.

Over this period, which exceeded one year, Musk's tweets and Tesla's announcements consistently demonstrated the measurement effect. Each public statement sparked significant price movements that extended beyond Bitcoin to the entire cryptocurrency market. It was like a quantum experiment with billions of dollars.

Quantum Finance Insights: Tesla's foray into Bitcoin highlights a phenomenon in finance that closely mirrors the quantum observer effect. Each time Tesla made its Bitcoin activities public, these announcements changed the state of the market itself. Just as observing a quantum particle influences its attributes, Tesla's statements about Bitcoin directly altered how the market valued and perceived the cryptocurrency.

In a metaphorical sense, this situation was like witnessing quantum entanglement within financial markets. In physics, measuring one entangled particle instantly determines the state of the other, irrespective of distance. In a similar manner, Tesla and Bitcoin became financially "entangled" due to shared market sentiment. News about Tesla's Bitcoin approach immediately impacted Bitcoin's price, and the effect was mirrored back all through swift shifts in market sentiment.

What was revealed in Bitcoin's price wasn't its intrinsic value, but rather its reaction to the "measurements" made by major players in the market.

Practical Application: Investors should incorporate quantum thinking and recognize that influential market participants don't merely observe markets. They can fundamentally alter the markets through their observations. This means your monitoring of the actions of significant players is not sufficient. You need to understand how those actions collapse probability distributions into new market realities.

Markets influence not only the fundamentals but also the way influencers share their stories. This quantum perspective acknowledges that in the interconnected financial system, observation and participation are inseparable aspects of market function.

Brexit referendum (2016)

The Quantum State: Pre-referendum, the UK economy existed in a superposition of two potential futures of continued EU membership or independence. Polling and projections functioned as measurements that themselves influenced market behavior and, ultimately, the vote itself.

The Story: On June 23, 2016, the United Kingdom held a referendum on whether to remain in the European Union or leave the bloc. This binary choice created a classic example of the measurement effect in real life. Polls, projections, and market reactions formed a complex feedback loop that influenced both financial markets and voter sentiment.

In the months before the referendum, financial markets exhibited classic quantum behavior. It oscillated between pricing in "Remain" and "Leave" outcomes based on the latest polling data. The British pound, in particular, became a barometer for referendum expectations, with each new poll triggering significant price movements.

The remarkably close race, with neither side maintaining a consistent lead, compounded this volatility. The GBP/USD rate hit a seven-year low of $1.38 in February 2016 when "Leave" polls strengthened. As the referendum day approached, there was a feeling in the financial markets, betting markets, and polling experts that "Remain" would eventually prevail. The pound's performance on June 23 reflected this confidence, rising to $1.50 against the dollar, its highest level in six months. The FTSE 100 also closed up 1.2% on referendum day. The banking stocks stood strong, signaling that the markets were expecting a "Remain" victory.

That was not the case, and "Leave" won with 51.9% of the vote, defying market expectations. This unexpected result triggered one of the most dramatic currency collapses in modern history. The pound fell by over 10% against the dollar overnight, reaching its lowest level since 1985. The FTSE 100 initially plunged 3.15%. The FTSE 250, which primarily focuses on domestic companies, experienced a decline of 11.7%. The UK bank stocks lost between £35 and £40 billion in value overnight. Barclays and

Royal Bank of Scotland both fell between 20% and 30% at the open, briefly triggering trading halts.

The market reaction itself became a factor in subsequent political developments. During the campaign, Leave campaigners had dismissed predictions of economic damage, but the severe market reaction reinforced them. In a quantum-like twist, the very act of measuring public opinion appeared to influence the outcome.

Post-referendum analyses suggested that polls predicting a "Remain" victory might have led to complacency among some "Remain" voters, reducing turnout. The same prediction motivated "Leave" supporters, though direct evidence is limited.

The Quantum Finance Lesson: The Brexit referendum demonstrated that markets react not only to actual events but also to attempts to measure their likelihood. Much like the famed double-slit experiment, where observing electrons changes their behavior, financial markets shifted in response to polling data that sought to predict voter intentions.

This example also highlights a kind of financial entanglement across different asset classes and global markets. The outcome of the referendum resonated not only with UK assets but also with international markets. The yen strengthened as investors sought safe havens, and European bank stocks experienced sharp declines. This swift, interconnected reaction resembled quantum entanglement, where the state of one market instantly affects others through shared market sentiment. In this way, Brexit serves as a powerful illustration of how financial systems can exist in a state of quantum uncertainty, with forecasts and predictions themselves shifting the probability landscape.

Practical Application: Investors should adopt quantum thinking by recognizing that market consensus itself becomes a factor in how events unfold. Even when positioning becomes too one-sided based on projected outcomes, that does not rule out the potential for quantum-like reversals. The lesson is maintaining scenario flexibility rather than making binary bets on political outcomes.

COVID-19 vaccine announcements

The Quantum State: Global financial markets existed in a state of superposition between pandemic-driven recession and recovery. Vaccine efficacy announcements functioned as measurement events that collapsed these possibilities into new market realities.

The Story: The COVID-19 pandemic created unprecedented uncertainty in global financial markets. By late 2020, economies worldwide had experienced severe contractions. The International Monetary Fund (IMF) estimated a 4.4% decline in global GDP for the year, which was the worst global recession since the Great Depression. Against this backdrop, vaccine development became the critical variable determining the timing and strength of economic recovery.

On November 9, 2020, Pfizer and BioNTech announced preliminary results from their Phase 3 vaccine trial, indicating an efficacy rate of over 90%. The market reaction was instantaneous and dramatic. The S&P 500 jumped 1.17% at the open, while the Dow Jones Industrial Average moved by over 800 points. Stocks of companies that supported work-from-home requirements, like Zoom and Peloton, plummeted. While Zoom fell by nearly 17%, previously battered sectors such as airlines, hotels, and oil companies recovered. In an emphatic demonstration of the measurement effect, billions in market capitalization shifted between sectors within hours.

Just one week later, on November 16, Moderna announced its vaccine showed 94.5% efficacy. Again, markets responded enthusiastically, though with less dramatic moves, as investors had already adjusted pricing for the reopening of the economy. Nevertheless, the Dow Jones Industrial Average rose 1.6%, and the Russell 2000 index of small-cap stocks, which is more sensitive to economic reopening, went up by 2.37% to a record high. The pattern repeated on November 23, when AstraZeneca announced its vaccine showed 70% efficacy overall, with one dosing regimen reaching 90%. AstraZeneca's vaccines had lower costs and simpler storage requirements, which were advantageous for global distribution.

These vaccine announcements also created financial entanglement across markets. Just as quantum particles correlate instantly, airline and oil stocks surged while tech stocks tied to remote work fell, showing how one sector's state instantly reshaped another's through shared market sentiment.

Perhaps most intriguing was how markets responded to subsequent vaccine news. On November 30, Moderna announced it would apply for emergency Food and Drug Administration (FDA) authorization after final analysis confirmed 94.1% efficacy. Interestingly, markets barely moved on this news. This diminishing market impact demonstrates how repeated measurements in quantum financial systems have less effect as uncertainty decreases. Each announcement reduced the range of possible economic outcomes, stabilizing Q$ asset valuations.

As vaccines began receiving regulatory approvals in December 2020, market focus shifted from efficacy to distribution challenges.

The Quantum Finance Lesson: The way financial markets responded to COVID-19 vaccine announcements perfectly demonstrates how these systems can mirror quantum principles. Before Pfizer's breakthrough, markets were balancing between multiple futures – ranging from prolonged economic downturn to swift recovery – much like a quantum superposition. When the efficacy results were made public, this uncertainty was instantly resolved, with market expectations adjusting right away. Each successive vaccine announcement acted as another measurement event, further reducing ambiguity and shaping new realities for investors. Unlike classical systems, where measurement simply reveals what is already true, in finance, these announcements actively shape and create new market outcomes.

This example underscores why it's vital to see financial markets as ever-changing systems of ongoing measurement and adaptation, rather than as static valuations. Understanding announcements as "events that collapse uncertainty" is key to grasping how markets move in response to major news.

Practical Application: Investors should adopt quantum thinking by recognizing that market-moving announcements don't simply reveal information. These announcements could fundamentally transform the probability landscape. This requires you to position portfolios to adapt to potential measurement events rather than making binary bets on specific outcomes. This quantum approach acknowledges that markets don't simply move from one equilibrium to another but exist in a constant probability flux.

Looking ahead

This chapter discussed how the very act of financial measurement reshapes reality, including Brexit polls influencing voting behavior or earnings reports driving business strategy. Every observation creates a feedback loop that amplifies outcomes far beyond what traditional models predict. But if measurement changes everything, how do we make decisions in such a fluid environment? The quantum nature of finance, beyond challenging how we observe markets, demands entirely new approaches to planning and strategy. In our next chapter, we'll explore how to build decision-making frameworks that thrive on uncertainty rather than being paralyzed by it. You'll discover why traditional budgets become straitjackets in quantum markets and how central banks navigated the impossible task of managing post-COVID inflation when every policy signal became a market-moving event. Get ready to transform your planning from static predictions into dynamic adaptation systems.

Chapter 5

Dynamic decision-making in a changing financial landscape

"*In a storm the wise bird sits on the flexible tree, never on the rigid oak*," goes an old Quantaland proverb.

Mariya, a handicraft supplier, has a year-long contract with a retailer. Just about a month after the contract, the global shipping crisis spiked all costs, and her finished goods piled up, stranding them and rendering her budget useless. Rigidity in her plan, which she felt was virtuous, failed her. This chapter equips you with frameworks for dynamic decision-making, showing how adaptive budgeting responds to market signals. You will understand that working capital thrives in disruptions and appreciate that machine learning (ML)–powered predictive analytics can help dynamic decision-making. You will study case studies like central banks' inflation responses and witness the transformation of financial planning from static exercises to a continuous adaptive process.

Beyond the snapshot: why financial plans must be live

In the quantum world, particles are never still. They exist as clouds of probability, only snapping into a fixed state the moment they are observed. Financial markets behave in exactly the same way. Market conditions, competition, and economic fundamentals are in a perpetual state of flux, turning yesterday's perfect strategy into an obsolete blueprint today.

This constant motion demands quantum-inspired financial planning that treats budgets as probability distributions rather than fixed targets. Traditional finance spends immense energy building elaborate long-term plans based on the hope that the world will stand still long enough to use them. Quantum finance rejects this hope. It builds flexible frameworks designed to expect change and adapt in real time. This isn't a rejection of long-term vision. It is a direct challenge to the fatal assumption of stability that dooms traditional plans from the start. The goal is to stop reacting to change and start anticipating it, to prepare for a spectrum of scenarios, and

DOI: 10.4324/9781003619635-5

to maintain the flexibility to pivot the instant the next measurement reveals a new reality.

Building for flux: the toolkit for adaptive strategy

Success in today's markets doesn't come from a perfect plan but from a plan that can change. You need strategies that evolve with new information, not ones that resist it. Adaptive approaches abandon the futile quest to predict the future. Instead, they build an organization that can respond effectively to the moment market conditions shift and reveal a new reality.

The fundamental distinction lies in philosophy. Traditional finance views uncertainty as a problem that requires improved forecasting. The quantum approach embraces uncertainty as a fundamental feature of the system. It uses flexible tools, such as rolling forecasts, to navigate this inherent unpredictability.

Dynamic budgeting and financial planning

The annual budget is a relic of the past that has served us well. But it locks you into assumptions that are often obsolete within months. Dynamic budgeting accepts that your financial plan exists in a superposition of possible scenarios. You only collapse it into specific allocations when market conditions force your hand.

Quantaland Retail does this with rolling forecasts. If customer behavior suddenly shifts from stores to online, their system instantly moves money from in-store operations to digital marketing. This creates a powerful competitive edge that a static budget can never match.

Here's the quantum insight. Your budget both reacts to the market and actively changes it. When Quantaland shifts its marketing expenditure, competitors notice and react. This creates a feedback loop, a form of financial entanglement, where your planning directly influences your competitors' actions. This ends up with all players collectively shaping the market's reality.

Zero-based budgeting is another adaptive tool. It forces every expense to justify itself based on today's reality, not last year's tradition. When a new payment platform slashes transaction costs for Quantaland Manufacturing, the savings aren't just banked; they are instantly reallocated to R&D.

Finally, scenario planning is the ultimate quantum budget. You don't pick one future. You actively prepare for multiple optimistic, realistic, and pessimistic outcomes at the same time. When the market finally collapses the uncertainty, you are ready to move instantly without scrambling to create a new plan.

Working capital optimization

In a dynamic world, working capital can't be managed on autopilot. Your inventory, receivables, and payment terms must be as fluid as the market itself. Adaptive strategies treat supply chains as unpredictable networks and manage cash and inventory across multiple scenarios.

A closure at a port in Quantaland is a perfect example. Adaptive companies view disruptions as new realities rather than temporary obstacles. They might boost strategic inventory, which in turn changes their relationship with suppliers, potentially locking them into new volume commitments. Your working capital decision doesn't just respond to the supply chain; it changes its very structure like entangled particles influencing each other.

This interconnectedness applies everywhere. The credit terms offered by Quantaland Services instantly affect their competitive position and cash flow, a connection that static policies miss. Even cash management needs to be prepared for significant changes. A laddered investment structure that is useless in a low-rate environment becomes essential when rates suddenly rise. Working capital is a tool for navigating uncertainty, not merely funding day-to-day operations.

Performance management system adaptation

You cannot establish inflexible performance metrics. Your KPIs must evolve as fast as your business model and the market change. Dynamic metrics reflect today's priorities, not yesterday's. This system sees metrics as existing in a superposition of potential uses, only becoming truly relevant when a business shift forces them into focus. A company doesn't progressively transition from efficiency metrics to innovation metrics; instead, it abruptly switches between them in response to competitive threats.

Incentive structures are entangled systems. A bonus plan drives individual performance along with affecting team collaboration and responding to market conditions. It holds multiple potential impacts until it collapses into a specific set of behaviors and outcomes.

When Quantaland Software shifted from selling licenses to subscriptions, they didn't just change their product; they changed their KPIs. They stopped measuring "deals closed" and started measuring "customer retention." This shift in measurement didn't just document the new strategy; it drove it. The sales team's behavior and their relationship with customers changed overnight. Real-time reporting is essential for this, capturing performance that is in constant flux instead of relying on stale monthly snapshots that seek false precision.

Risk-management framework evolution

Static risk management is a dangerous fantasy. A modern framework understands that risks exist in a superposition of potential impacts. They only collapse into a specific reality when a triggering event occurs. This requires a dynamic system that adapts to evolving threats and maps their interconnections.

The key insight is that risks are entangled. Supply-chain disruption creates cascading effects, instantly generating operational and financial risks beyond the initial logistics problem. Static risk matrices, which treat risks as isolated items, are constantly blindsided by these cascading effects.

At Quantaland Bank, stress testing isn't an annual exercise. It's a continuous process where updating scenarios actually influences the risk landscape itself. Monitoring alters managers' decisions, demonstrating an observer effect that modifies the very risks under measurement. This builds a quantum mindset, preparing the bank for a range of possibilities, not just one.

This also applies to insurance. A cyber policy can be considered both adequate and inadequate until a breach occurs, which eliminates that uncertainty. Business continuity plans must account for quantum-leaping operations between centralized and distributed models. Risk management must embrace this uncertainty, rather than assuming a static world.

Strategic finance decision-making

Strategic finance can't be about single-point estimates. It must account for the fact that a project's value exists in a superposition of possible outcomes until the market forces it into a single reality. Adaptive frameworks treat inputs like growth rates as evolving probabilities, creating dynamic valuations that make static DCF models look useless.

This is where real options create immense value. A capital investment isn't a yes/no decision. It's a decision that maintains multiple options: expand, pause, or exit. You only collapse these options into a specific path when new market data arrive. Real options valuation both measures flexibility and creates a quantum mindset that influences resource-allocation decisions.

This thinking also transforms M&A. A target's value is entangled with integration success, market shifts, and competitor reactions. Static valuation multiples completely miss this web of dynamic interdependencies.

Even the act of valuation itself is a quantum event. When Quantaland Energy evaluates a renewable project, the very act of measurement triggers strategic adjustments that change the project's future cash flow and thus its value. Valuation doesn't just assess value; it helps create it.

This is very evident in international expansion. A market's attractiveness exists in multiple states. A sudden trade war or currency crash doesn't gradually alter conditions; instead, it abruptly shifts the market from a favorable to an unfavorable state, necessitating an immediate adaptive response. Strategy is no longer a five-year plan; it is a constant state of prepared agility.

Powering decisions with real-time data

Advanced analytics provides the tools for dynamic decision-making in today's complex markets. Real-time data processing, automation, and predictive modeling let you spot changes as they happen and respond instantly. But this technology is useless without the organizational will to adapt and act. True success comes from marrying powerful analytics with a culture that promotes and empowers rapid response.

This requires a fundamental mindset shift. Instead of managing change as an exception, you must embrace it as the default state of your environment. Real-time analytics gives you this power: the ability to make decisions as the world evolves around you, not months later when a quarterly report tells you what already happened and is irrevocable.

Financial reporting and analytics infrastructure

Real-time analytics turns your financial reporting from a historical record into a live nervous system. It adapts to new realities instantly, unlike the static snapshot of a period-end close. This is quantum decision-making in action. Your performance metrics exist in a superposition of possibilities until a management decision locks in a specific outcome. Your financial data are a temporary state, constantly adjusting to new information.

Quantaland Chemical shows how this works. Their automated systems detect a spike in raw material costs and instantly trigger price adjustments for customers. The reporting of both documents actively shapes performance. Measuring alters the very reality under scrutiny.

This transforms variance analysis too. A threshold breach isn't a simple alarm. It exists in a superposition of potential urgency levels until management attention collapses it into a specific priority. Exception reports both identify problems and immediately direct organizational focus, shaping operational reality on the fly. This approach accepts variance as a normal part of the system, not an error to be corrected. Your metrics adapt to new realities; they don't try to force reality back to an old "normal."

Cash flow and liquidity management

In a dynamic world, your liquidity is never a single number. It exists in the superposition of possible states until an operational decision, like a large payment or receipt, forces it into a specific position. Real-time systems track cash flow as a set of evolving probabilities, not a fixed bank balance.

At Quantaland Transportation, this is a continuous loop. Their real-time fleet monitoring doesn't just track fuel costs and maintenance needs; it triggers immediate operational changes that directly reshape future cash flow. The forecast predicts liquidity while actively helping to create it through feedback loops.

This dynamic management optimizes the timing of every cash movement. A decision to adjust inventory levels doesn't exist in a vacuum. It instantly changes payment cycles to suppliers, storage costs, and ultimately customer service levels. These are connections that traditional cash management completely overlooks.

This approach treats liquidity data as a temporary quantum state – always ready to adapt to a new reality. You need systems built for constant change, not ones that assume a static environment.

Credit and risk analytics

Dynamic credit risk perceives a customer's creditworthiness as a vast array of potential outcomes. This quantum state collapses into a fixed "good" or "bad" only when actual payment behavior is observed. Real-time analytics treats risk as being in constant flux, unlike static quarterly reviews that pretend risk is a fixed fact.

This changes everything about collections. A payment reminder isn't just a response to a late payment; it's an intervention that actively alters the customer's future payment behavior. This creates powerful opportunities to prevent problems before they happen rather than just writing them off after the fact.

Quantaland Equipment Finance lives this way. Their real-time monitoring of payment patterns changes risk while being in the act of assessing the same. By adjusting collection efforts based on live data, they directly influence customer behavior. The act of measuring risk becomes part of managing it.

This approach embraces uncertainty instead of fighting it. Credit relationships are probability distributions that never stop evolving. Consequently, you need systems that adapt to this constant change, not ones that try to predict a fixed future.

Regulatory and compliance technology

Automated compliance doesn't just monitor your regulatory status; it actively transforms it. This is dynamic decision-making, where compliance exists in a superposition of possible states until a regulator's examination forces a specific outcome. Real-time systems adapt to evolving rules, unlike static checklists.

At Quantaland Bank, automated reporting creates a powerful observer effect. Real-time monitoring identifies problems while also triggering immediate corrective actions that change the bank's compliance reality. The system prevents violations instead of just documenting them, drastically reducing workload and risk.

This interconnectedness is critical. Anti-money laundering systems don't look at transactions in isolation. They see the patterns across multiple transactions. A single suspicious transaction changes the risk score of every related transaction, a connection that manual reviews mostly miss.

This quantum thinking applies perfectly to tax. The more precisely you set a transfer price today to fit "arm's-length" rules, the harder it is to defend that exact price in a future audit when market conditions have inevitably changed. Tax obligations exist as a probability distribution until the moment of regulatory examination.

Continuous audit creates a wonderful feedback loop. The testing process itself influences the controls being tested, enabling real-time adjustments that maintain compliance as conditions change. This technological foundation finally allows finance professionals to operate in the present, making decisions as the world evolves, not reacting to yesterday's news.

The great divide: static maps versus dynamic GPS

Think of the gap between traditional financial models and market reality as the difference between a photograph and a live video broadcast. Traditional models are detailed snapshots of the past, built on the assumption that historical conditions will persist. Dynamic models function as a live feed, continuously adapting to new data as circumstances change. This section will show you why relying on those outdated snapshots creates dangerous blind spots and how a dynamic approach bridges the gap.

But you must see this as more than a technology upgrade. The core challenge isn't just about getting newer data faster. It demands a fundamental mindset shift in how you approach decision-making itself. Traditional models seek to eliminate uncertainty through exhaustive analysis of historical patterns. Dynamic models, like quantum systems, embrace uncertainty. They are designed to adapt and refine your decisions as new data emerge, turning flux into a strategic advantage.

Stale data risks

Are you sure you're not making critical decisions based on outdated information? This inherent time lag isn't a minor inconvenience; it engineers systematic blind spots into your process. These blind spots lead to poor capital allocation, missed opportunities, and unexpected losses. The danger intensifies when your organization treats historical data as predictive truth, failing to recognize its inherent limitations.

Historical financial analysis limitations

Dynamic analytics outperforms static ratio analysis by adapting to current conditions, avoiding misleading time-stamped values. Financial statements oversimplify complex business realities into fixed snapshots. However, these realities hold multiple possibilities, like quantum states, until management decisions lock in actions. Financial reporting lag plays a major role in creating decision-making challenges. The more precisely you measure historical performance, the less accurately you can assess current conditions. Accounting measurements capture the continuous flux of business fundamentals at specific points, but they may already be outdated when decision-making takes place.

Let us look at another aspect of historical analysis. When Quantaland Retail moved to online sales during the pandemic, quarterly inventory turnover ratios generated conflicting signals. Traditional metrics showed improved efficiency achieved through order-driven procurement while hiding obsolete physical stock. The measurement method determined which aspects of reality were visible for decision-making, but it overlooked the fundamental transition between business models.

Industry benchmarks illustrate different analytical perspectives on performance measurement. When viewed as trends over time, they show smooth patterns. The same data reviewed as discrete comparison points become snapshots that may not capture shifts in market conditions. Pre-cloud technology benchmarks miss shifts to improved technology. When the e-commerce boom changed customer behavior and competition, these benchmarks proved inadequate.

Budgeting and forecasting obsolescence

You must stop treating your budget as a single, fixed projection of the future. A traditional budget is a fantasy of stability. Dynamic budgeting treats the future as it actually is – a probability distribution across multiple potential outcomes. This creates immediate constraints for you, as market conditions now shift faster than what annual budgeting and quarterly planning cycles can possibly accommodate.

Unlike rigid long-term budgets, dynamic budgeting adapts to market uncertainty using real-time data. Your budgeted assumptions don't gradually evolve. They can shift suddenly between different states, rendering meticulous budget precision meaningless for executives who need to make dynamic decisions. Often, the planning process itself becomes the most significant constraint, where teams optimize for hitting budget targets rather than responding to actual market dynamics.

The five-year strategic plan at Quantaland Media is a classic example of this planning obsolescence. Their plan assumed the continued dominance of traditional advertising while digital transformation was actively reshaping the entire industry. It treated competitive conditions as static – completely blind to the rapid transitions occurring in customer behavior and technology adoption. The plan was a detailed map of a world that no longer existed.

This applies directly to capital allocation. Basing decisions primarily on historical returns reflects a deep limitation. Here, a crucial irony emerges: the more perfectly you model past performance patterns, the less accurately you may predict future market changes. Even your investments can yield multiple potential returns. The investment exists in a superposition of possible returns until market forces cause it to collapse into a single outcome, which may shatter all your historical precedents.

Your resource planning demonstrates these interconnected challenges. Historical demand assumptions became completely obsolete during recent global supply-chain shifts, where exceptional conditions revealed hidden interdependencies that old models couldn't see. Traditional planning tools are simply not equipped to anticipate, let alone accommodate, such rapid and fundamental changes.

Risk assessment and management gaps

Be careful; your risk models may be trying to solve the wrong problem. They mostly attempt to treat risk as a fixed entity with stable correlations. In reality, risk is a quantum system. Correlations hold multiple possibilities until an intervention, like the act of observation, forces them into a single, often terrifying, reality.

The 2008 financial crisis was not a statistical outlier. It was a measurement.

It revealed the hidden entanglement of assets that classical models had assumed were independent. The risk was always there, existing in multiple states; the crisis simply collapsed the system into its worst possible configuration.

Please review your credit scoring models. They assess risk, and at the same time, they create the behavior they measure. Pre-pandemic models were built on assumptions of stable employment and economic relationships. The pandemic was a massive, unforeseen observation that collapsed those assumptions, making traditional debt-to-income ratios temporarily irrelevant and revealing hidden connections the models could never see.

The same is true for operational risk. The more precisely you model past threats, the more vulnerable you become to novel ones that defy historical patterns. Pre-pandemic cybersecurity models, built for a centralized office environment, were based on the assumption of a static state. The sudden shift to remote work was a quantum leap that collapsed their defensive posture into a state of inadequacy overnight, leading to breaches from unforeseen vectors.

Your VaR calculations can create a dangerous illusion of security. They treat extreme market events as statistical outliers. But a crash isn't an outlier; it's a phase transition where the market leaps into a new state, making all historical probability distributions useless.

Audit and assurance challenges

You must understand that an audit changes the very system it examines. This is the observer effect in action within your control environment. Traditional auditing assumes a stable system. A dynamic audit recognizes that controls exist in a superposition of states, namely adequate and inadequate, until your testing forces a specific result.

At Quantaland Systems, during their digital transformation, every control existed in two states at once. How did we know if a control was effective? It depended entirely on whether you measured it against the legacy system or the new digital requirements. Only the act of auditing itself could collapse that uncertainty into a definitive answer.

Your internal control testing creates its own reality. The audit process itself can influence control effectiveness and subsequent management decisions. Controls exist between adequate and deficient states until testing reveals their current reality. Operating models can shift into irrelevance while new technologies create control gaps completely unknown to traditional testing approaches.

Your materiality assessments are a powerful example of shaping reality. By focusing too precisely on historical benchmarks and ratios, you can reduce your accuracy in identifying emerging, non-financial risks. As business models evolve, stakeholder priorities exist in flux until your audit decisions force specific testing procedures. This measurement can prompt a leap in the business significance of a model, abruptly shifting decision-making priorities across the organization.

Strategic decision-making delays

You need to start seeing market opportunities as existing in a quantum state. They are clouds of potential until a competitor makes a move, or you act, collapsing them into a single, captured reality. Your strategy can no longer be a five-year map. It must function as a dynamic positioning system that updates in real time.

This thinking applies to organizational restructuring. You face a painful irony where the more efficiently you optimize your current organizational structure, the more you may be locking yourself into the past. Your efficiency metrics are in a superposition of being relevant and obsolete until a sudden market shift forces a measurement and may necessitate a rapid reorganization.

Quantaland Consumer Products experienced this with their launch into the wellness market. Consumer preferences were in a state of flux, in a superposition of possible trends. Their extensive market research was an attempt to measure and collapse that uncertainty. But the act of research took time, and during that observation period, consumer preferences leaped again. They launched a product for a reality that had already vanished, leading to a costly misalignment.

Finally, your product development, too, operates in an entangled state. There is a connection between customer needs, technological capabilities, and competitive moves. Historical preference data are a probability distribution that can abruptly become irrelevant. There were technology companies that were still optimizing pre-remote-work software requirements. Their product roadmap underwent a complete overhaul once they realized that they were building for a world that had already undergone measurement-induced change.

The adaptive mind: machine learning for quantum finance

AI and ML provide powerful tools for dynamic decision-making that can adapt to new patterns more effectively than static models. ML identifies subtle patterns, updates its assumptions automatically, and uses those for prediction. The key advantage lies not just in computational power but in its ability to factor in new information while making decisions.

The quantum-like insight is the key. ML models don't just predict outcomes. They create decision frameworks that can recognize when fundamental assumptions have changed and adapt accordingly. This represents a fundamental shift from trying to predict the future to building systems that can respond effectively when it reveals itself.

Predictive financial analytics

ML systems outperform static forecasts by treating financial variables as probabilities, not certainties. Where traditional forecasting forces revenue into deterministic projections, ML models see revenue as existing in a superposition of possibilities. They adapt dynamically as market conditions evolve, collapsing these possibilities into a specific forecast only when necessary for a decision.

At Quantaland Technology, their dynamic revenue models process subscription renewals, usage patterns, and economic indicators in parallel. Unlike static models that cram multiple variables into a single forecast, the ML system maintains flexibility across potential outcomes. This doesn't give you one answer; it enables your decision-makers to prepare for multiple scenarios, shifting resources as probabilities change.

This extends to cash-flow prediction. AI-powered tools create an interconnected decision-support system where operational indicators and liquidity states influence each other. Production schedules, payment terms, and supplier data connect to form a probability field, not merely a set of isolated values. When one element changes, for instance, a delayed shipment, the model instantly recalculates its impact on the entire cash-flow ecosystem, proactively triggering appropriate liquidity decisions.

Risk management and fraud detection

ML-based risk models fundamentally redefine how you classify threats. Traditional rule-based systems see fraud in binary terms: yes or no. Dynamic systems, like quantum states, see fraud as a cloud of possibilities, adapting the level of suspicion as transaction patterns evolve.

Quantaland Finance's fraud detection exemplifies this. Their AI tool observes transactions that, at the same time, exist as both legitimate and suspicious until classification occurs. The act of algorithmic observation, like analyzing patterns in real time, resolves this uncertainty. These models, apart from detecting known fraud, enable proactive decision-making by recognizing emerging patterns before they crystallize into actual losses.

This creates a powerful feedback loop. The very act of detection shapes criminal adaptation strategies. Your security teams, armed with this knowledge, can now stay ahead of evolving threats, making strategic decisions to harden defenses against the next attacks, not just last year's.

Customer and market analytics

Forget static customer segmentation. AI-driven analytics recognizes that customers exist in probability distributions that are continuously evolving. Thus, customer value isn't a fixed number. Instead, it's a set of multiple possibilities that collapse into a specific outcome only upon a key behavioral event, like a purchase or churn.

This creates a powerful observer effect. When Quantaland Airlines identifies a customer as high-value and targets them with premier loyalty offers, that very classification changes the customer's pattern of engagement. The measurement alters the measured. This makes customer valuation an active process of shaping reality beyond passively documenting it.

You see this in the pricing at Quantaland Hotels. Their dynamic pricing algorithm links room rates to demand, weather, and competitor actions, forming a live correlation network. These variables exist as a probability field, shifting between different pricing regimes until market conditions force a specific rate decision. This interconnected view is far more likely to maximize revenue and occupancy than analyzing each parameter in isolation.

Investment and capital allocation

ML-driven portfolio optimization accepts the fundamental truth that asset correlations are not stable. They exist across multiple potential states until market stress or a major event forces them into a specific pattern. AI-driven tools adapt to these shifting correlations in real time, moving far beyond the static assumptions of mean-variance models.

This quantum thinking applies to corporate finance. At Quantaland, their AI-powered M&A analysis demonstrates a clear observer effect. The algorithms used to screen and value acquisition targets inevitably influence market dynamics and can even shape the behavior of the target companies themselves. This makes valuation an active process of decision-driven reality creation, not just the passive discovery of a predetermined number.

The approach supports more informed decisions about timing, pricing, and integration strategies based on dynamic market conditions. The decisions are more responsive to changing environments than those made using static financial multiples.

Regulatory and compliance automation

Static, rule-based compliance systems are brittle. AI-driven compliance adapts to evolving regulations and enforcement interpretations, which exist in a state of flux like quantum states. It takes a regulatory examination to collapse them into a specific ruling.

The real power lies in mapping the hidden interdependencies. An AI system at Quantaland Healthcare doesn't assess privacy laws in isolation. It analyzes how a change in one regulation connects to data security protocols, marketing policies, and operational procedures, forming a complex compliance network. This enables your team to make dynamic decisions about policies and procedures based on the evolving regulatory interpretation, not just the black-letter law. This keeps you ahead of the compliance curve.

This approach is more robust and secure than anything achievable through static rule interpretation. It transforms compliance from a historical documentation exercise into a forward-looking strategic function.

Meanwhile, far from Quantaland

Central banks' responses to inflation post-COVID

The Quantum State: Central banks operated in a state of probabilistic decision-making, where inflation existed as both transitory and persistent until data resolved the uncertainty. Economic data eventually collapsed these possibilities into clearer realities.

The Story: The post-COVID inflation surge presented central banks worldwide with one of their most challenging decision-making environments. As economies reopened in 2021, inflation rates in developed economies surged to higher levels, forcing monetary authorities to continuously reassess their policies in real time.

Initially, most central bankers characterized rising inflation as a temporary phenomenon resulting from supply-chain disruptions and pent-up demand. They also attributed the surge to the base effects of 2020's depressed prices. Most economies expected that the increases in prices are a one-time feature and are likely to have only transient effects on inflation.

However, as 2021 progressed, inflation proved more persistent and widespread than initially expected. U.S. consumer prices rose 7% in 2021, marking the largest annual increase since 1982. Eurozone inflation reached 5% by December 2021, while UK inflation hit 5.4% by January 2022. This forced central banks to rapidly adapt their assessments and policies.

The Federal Reserve's pivot was particularly dramatic. In June 2021, the Fed indicated no rate hikes until 2023. By December 2021, it projected three rate hikes in 2022. By March 2022, the Fed had begun the most aggressive tightening cycle since the early 1980s. They raised rates by over 500 basis points in just over a year. Jerome Powell, the Chair of the Federal Reserve, used the term "transitory," which took on new meaning as inflation persisted.

The Bank of England moved even faster, becoming the first major central bank to raise rates in December 2021. Though the ECB initially maintained an ultra-careful stance, it eventually joined the tightening cycle in July 2022. These differing timelines reflected varying assessments of inflation probabilities across economies. You can spot a classic example of quantum-like decision-making under uncertainty.

It is fascinating to note how the central banks adapted their policy frameworks to this uncertainty. The Fed's shift to average inflation targeting in 2020 was designed to allow inflation to run above 2% for some time to make up for previous shortfalls. But as inflation surged, the Fed was forced to reinterpret this framework dynamically. Recognizing the hardships induced by high inflation, the Fed prioritized restoring price stability while preserving a strong labor market.

This dynamic decision-making extended to communication strategies as well. Central banks increasingly emphasized data dependency rather than forward guidance, acknowledging the inherent uncertainty of the economic environment. The central banks maintained a more open and responsive mindset as they drafted successive monetary policies.

The Quantum Finance Lesson: The quantum nature of economic decision-making became clear in central banks' responses to inflation. Their strategies didn't follow a fixed path – instead, they shifted as new economic data emerged. Just as quantum particles are described by probabilities rather than set trajectories, inflation and related variables forced central bankers to continuously update their outlooks, reflecting a landscape of ever-changing probabilities.

This process highlights a quantum-inspired approach to navigating uncertainty. Central banks refined their inflation forecasts and policy decisions with every new piece of information, echoing how quantum systems only reveal their state once observed.

Ultimately, central bankers faced the ongoing challenge of distinguishing meaningful signals from noise within limited data, adapting their policies in real time as the broader economic situation became clearer.

Practical Application: Financial decision-makers facing evolving probability distributions should embrace quantum thinking by maintaining adaptability. This means avoiding rigid commitments to specific economic narratives and instead maintaining flexibility to adjust as new data emerge.

This quantum perspective acknowledges that economic decision-making isn't about determining a single correct path. Like the path of quantum particles, it involves navigating a constantly evolving probability landscape.

Hedge funds using algorithms: the case of Renaissance Technologies

The Quantum State: Renaissance Technologies' algorithmic approach treats markets as probability distributions rather than deterministic systems. These algorithms continuously adapt to evolving patterns without fixed assumptions about market behavior.

The Story: Renaissance Technologies, founded by mathematician James Simons in 1982, is a wonderful example of dynamic decision-making in finance. The firm's flagship Medallion Fund was known to generate average annual returns of approximately 66% (before fees) over more than three decades.

What sets Renaissance apart is its approach to markets as complex adaptive systems best understood through probability rather than causality. Unlike traditional fundamental investors who seek to understand why markets move, Renaissance focuses entirely on detecting statistical patterns in market data, regardless of their apparent economic logic. When prices go up, it is because the number of people wanting to buy exceeds the number of people wanting to sell. While most researchers sought to understand the reasons behind this phenomenon, Renaissance accepted it as a fact and investigated potential profit opportunities arising from it.

Simons had explained in a 2016 interview: "We're right 50.75 percent of the time . . . but we're 100 percent right 50.75 percent of the time. You can make billions that way." This probabilistic mindset embraces the quantum nature of markets. It acknowledges that perfect prediction is impossible, but that slight statistical edges can be enormously profitable when applied consistently across thousands of instances.

Renaissance employs a large number of doctorates in mathematics, physics, statistics, and computer science. They were drawn from diverse backgrounds, including cryptography, speech recognition, and computational linguistics rather than finance. This interdisciplinary approach allows the firm to detect patterns that traditional financial analysts might miss. They benefited from trying to analyze a problem from radically different viewpoints.

The firm's algorithms continuously analyze vast amounts of data. These data came from various sources. The sources include price movements, volume patterns, macroeconomic statistics, and even textual information from news sources. Their prime focus is to identify temporary market inefficiencies. Crucially, these algorithms are designed to adapt in real time to changing market conditions, modifying their trading signals as correlations shift. This dynamic adaptability was particularly evident during the COVID-19 market turbulence of March 2020. While many quantitative funds suffered significant losses as historical correlations broke down, Medallion reportedly generated positive returns. This underlined their ability to rapidly adjust to new market realities.

Renaissance's approach is notoriously secretive, with employees signing strict non-disclosure agreements. However, occasional glimpses into their methodology reveal their quantum-like characteristics. Their approach reflects a dynamic state, adapting to new market conditions. The short- to medium-term focus of Renaissance allows them the flexibility of adapting to the new reality.

The Quantum Finance Lesson: Renaissance's approach demonstrates the effectiveness of quantum thinking in financial markets. Rather than seeing price movements as predetermined, they treated them as probabilities and adapted to evolving market conditions. Much like quantum mechanics replaces certainty with probability, Renaissance shifted from believing in market predictability to focusing on recognizing statistical patterns.

This strategy embodies quantum-like thinking, where using various analytical models reveals different market patterns. Renaissance utilized multiple models at once, each highlighting distinct aspects of the market. They intentionally avoided relying too heavily on any single model, recognizing that no single method could capture the full complexity of the markets.

Renaissance understood that the patterns observed in the market are not absolute truths, but rather statistical relationships that depend on the analytical frameworks used. The insights gained about the markets always depend on the chosen method of analysis.

Practical Application: You should incorporate quantum thinking and recognize that markets exist as probability distributions rather than fixed systems. This requires focusing on statistical edges rather than perfect predictions to understand movements. You also need to be adaptable to evolving market conditions and use diverse analytical frameworks to capture complementary aspects of market behavior.

This quantum perspective acknowledges that markets, like quantum particles, cannot be perfectly predicted. You must identify the market state frequently and earn more from correct identifications than you lose from mistakes. There is no defined state where you can make profitable decisions; it is only when you define the state correctly that you can make them.

Walmart: inventory management during the pandemic

The Quantum State: Walmart maintained multiple potential supply-chain configurations in parallel, allowing rapid adaptation as pandemic conditions evolved and consumer behavior oscillated between competing states.

The Story: Walmart's response to the COVID-19 pandemic represents a successful example of dynamic decision-making. As the world's largest retailer, Walmart faced unprecedented challenges when global supply chains collapsed amid pandemic lockdowns. Furthermore, Walmart experienced an almost instantaneous transformation in consumer behavior.

Pre-pandemic, Walmart had spent years optimizing its supply chain for efficiency, employing just-in-time inventory management to minimize costs. This approach relied on stable demand patterns and predictable lead times. These assumptions were shattered in March 2020 when consumer panic-buying led to empty shelves. Factory closures, which disrupted product availability, exacerbated the problem.

Walmart's response was remarkable for its speed and adaptability. Instead of clinging to its established supply-chain model, the company rapidly reconfigured its entire system to prioritize availability over efficiency.

The company replaced its certainty-based planning with contingency-based planning. This transformation reflected a need for time.

Walmart moved away from lean inventory management practices to improve resilience. They increased inventory levels for essential products significantly in Q2 2020 compared to pre-pandemic levels. At the same time, they rapidly expanded their supplier base, bringing on new vendors to reduce dependencies on single sources. Specific focus was on critical products such as cleaning supplies and paper goods, which saw an unprecedented increase in demand.

Another remarkable move was omnichannel adaptation. Walmart accelerated its e-commerce capabilities. It expanded the number of online grocery pickup stores and deployed ship-from-store capabilities to utilize store inventory for online orders.

Walmart leveraged its vast data analytics capabilities to identify emerging consumer trends in near real time and adjust procurement accordingly. Perhaps most impressive was Walmart's dynamic response to the constantly shifting pandemic environment. When initial lockdowns created surges in grocery and essential goods, Walmart rapidly reconfigured store layouts and staffing. As shopping patterns evolved again during reopening phases, the company adjusted once more, restoring capacity in previously deprioritized departments like apparel.

This adaptability extended to consumer experience as well. Walmart quickly implemented over 150 safety-related changes to its operations. These included contactless payment options, enhanced cleaning protocols, and modified store hours. As consumer comfort levels and regulatory requirements changed, these measures were continuously adjusted rather than remaining static.

By the end of 2020, these dynamic adaptations had paid off significantly. While many retailers struggled with supply-chain disruptions, Walmart reported comparable store sales growth of 8.6% for fiscal 2021, with e-commerce sales growing 79%. Rapid adaptation to customer needs gave Walmart a competitive edge.

The Quantum Finance Lesson: Walmart's approach during the pandemic demonstrates how supply-chain management can resemble quantum-like behavior. Success depended on the ability to hold several different supply-chain models in readiness, allowing for rapid shifts as the situation evolved. Like quantum systems that exist in multiple possible states, Walmart's supply chain maintained a spectrum of options, prepared to adjust instantly as new circumstances arose.

This strategy mirrors the quantum principle of superposition, where elements can be in several states simultaneously until observed. By treating its supply chain as a suite of flexible capabilities rather than a fixed structure, Walmart was able to reconfigure its resources swiftly in response to changing market demands. This approach required an active commitment to ongoing evolution rather than relying on static, established processes.

Practical Application: You can incorporate quantum-like thinking by maintaining operational flexibility through deliberately redundant capabilities. This means avoiding overoptimization for a single business environment and developing the ability to rapidly reconfigure operations as conditions change.

This quantum perspective acknowledges that business environments, like quantum systems, exist in states of probability rather than certainty. Organizations need to maintain adaptability rather than rigidity to cope with the challenges of change.

Looking ahead

We've now built the foundation for dynamic decision-making, from adaptive budgets that respond to market signals to ML systems that treat uncertainty as fuel rather than friction. But there's a deeper quantum reality that makes this dynamic approach not just useful but essential: the profound interconnectedness of modern financial markets. When the Federal Reserve pivoted on inflation, it didn't just change U.S. policy. It instantly reshaped global bond markets, currency valuations, and investment flows worldwide. In our next chapter, we'll explore this quantum entanglement effect, where seemingly independent financial institutions and markets are so deeply connected that a mortgage default in Florida can threaten a bank in Frankfurt within hours. You'll discover why diversification, the bedrock of traditional risk management, may be an illusion in our quantum-entangled financial system, and how understanding these hidden connections can transform your approach to portfolio management and strategic planning.

Chapter 6

Entanglement: the interconnectedness of financial markets

"*In a spider's web, no thread stands alone*," goes an old Quantaland proverb.

Owner of a small business, Rick, saw his export deals collapse when a foreign bank failed. The loss rippled through global supply chains that he never knew connected him. He was blindsided by hidden ties. This chapter takes you through the web of financial interconnectedness, revealing how quantum entanglement explains rapid crisis spread. You will explore examples of contagion, examine network vulnerabilities, and analyze how digital technology accelerates these effects. The study will help you understand why traditional risk models fail. It will also show you how quantum thinking improves systemic risk assessment and portfolio management.

The web of fate: when your risk becomes my risk

Financial entanglement describes a state where institutions become so deeply interconnected that the fate of one instantly and inevitably influences the fate of others. In this condition, traditional buffers like geographic distance or seemingly unrelated business models become irrelevant. The financial system begins to behave less like a network of independent entities and more like a set of quantum-entangled particles, correlated in ways that defy classical logic.

This reality challenges a fundamental principle of traditional finance. We are taught that diversification across institutions, regions, and asset classes is the bedrock of risk mitigation. Yet quantum finance reveals this protection can actually be a dangerous illusion. In a crisis, these supposedly diversified and independent entities suddenly move together. They become entangled, behaving as a unified system that conventional risk models, built on assumptions of independence, are blind to.

This is why modern financial crises exhibit such terrifying speed and scale. Understanding financial entanglement explains how trouble in one obscure corner of the market can instantly propagate through the entire

DOI: 10.4324/9781003619635-6

global system, overwhelming defenses and consistently surpassing the worst-case scenarios envisioned by risk managers. Attempts to contain the problem often fail because solutions, such as bailouts or policy shifts, ripple through the entangled network, causing unintended and immediate consequences elsewhere. The problem goes beyond mere spreading; it manifests everywhere at once, dramatically increasing its scope and complexity.

The channels of contagion: how institutions entangle

Financial institutions don't just connect; they entangle. They exist in complex webs of relationships that create precisely the quantum-like effects described in the previous section. This entanglement operates through direct lending, derivative contracts, payment systems, and shared exposures to common assets. What follows is an examination of its most critical channels.

Banking networks and correspondent relationships

The global banking system is a primary vector for financial entanglement, with correspondent relationships serving as instantaneous channels for contagion. When Quantaland International Bank maintains accounts with correspondents in New York, London, and Tokyo, their financial health becomes quantum entangled.

This isn't merely about facilitating payments. It means that a liquidity problem at Quantaland International Bank immediately increases the counterparty risk for its correspondents worldwide, even those with no direct exposure to Quantaland's domestic market. The entanglement is bidirectional. Should a major correspondent bank in London face difficulties, Quantaland International Bank loses its lifeline to international payment systems and foreign exchange (FX) markets, triggering immediate operational and liquidity stress at home. Geographic boundaries offer no protection.

This principle extends to central banking. Institutions reliant on central bank funding become entangled with monetary policy decisions. A shift in policy rates or liquidity facilities doesn't ripple slowly through the system; it propagates instantly, simultaneously altering the state of every connected bank, much like a quantum measurement collapsing a particle's state.

The 2008 crisis provided a quintessential example. The interbank lending market, which serves as the circulatory system, experienced an immediate freeze. Uncertainty about counterparty risk created quantum-like correlations, simultaneously making every bank a potential failure candidate. For Quantaland International Bank, this meant correspondent banks abruptly halting trade finance facilities, amplifying a local problem into a systemic threat overnight.

Insurance and reinsurance web

The insurance industry creates complex interconnections through its global reinsurance networks. When Quantaland Insurance writes a policy for earthquake coverage, it doesn't retain the risk. Quantaland Insurance distributes the risk to reinsurers in London, Bermuda, and Zurich, forming a network that inherently links each participant's fate.

A major catastrophe acts as a massive quantum measurement event. The 2011 Japanese tsunami didn't just impact Japanese insurers; it simultaneously affected balance sheets across the entire global reinsurance network, revealing correlations that were invisible during calm markets.

This creates dangerous "reinsurance spirals." Quantaland Insurance might cede risk to a Bermuda reinsurer, who then retrocedes a portion of that same risk to a firm in London. The London firm may already be exposed to the original Quantaland earthquake risk through other contracts. These hidden overlaps within financial networks intensify losses once a triggering event occurs, exposing previously concealed concentrations of risk. Only when a specific loss scenario forces the system to react do such vulnerabilities become evident, revealing the true extent of interconnectedness.

Lloyd's of London syndicates exemplify this. While each syndicate appears to operate independently, their participation in similar risks through different channels creates powerful entanglement. They remain correlated particles until a major loss event measures the system and reveals its true interconnected nature.

Investment fund interconnection

Modern investment management entangles institutions through shared holdings, homogeneous strategies, and common funding sources. When pension funds, mutual funds, and hedge funds hold similar assets, they become entangled, predisposed to generate simultaneous, massive selling pressure during market stress.

Exchange-Traded Funds (ETFs) create particularly potent entanglement. Their redemption mechanism means that selling pressure on the ETF share instantly forces the liquidation of the underlying securities. This selling then impacts the value of all other funds that hold those same assets, causing a rapid transmission of stress throughout the system.

The rise of passive investing has created a new form of structural entanglement. When a stock like Quantaland Motors enters a major index,

index funds are compelled to simultaneously buy billions of dollars of its shares. This boosts the price independent of the company's fundamentals, entangling the fates of disparate companies simply because they share a line in an index.

This extends to alternative investments. Hedge funds pursuing similar strategies, whether quantitative, volatility-based, or macro, become entangled through their exposure to common risk factors. In a crisis, they all experience correlated losses and margin calls despite holding different portfolios, creating systemic vulnerabilities with seemingly diversified strategies.

Cross-border financial dependence

Entanglement effortlessly transcends national borders. When multinational companies like Quantaland Manufacturing borrow in U.S. dollars to finance their European operations, they immediately entangle Quantaland's financial system with U.S. credit markets and European business circumstances.

Currency markets exhibit near-pure entanglement effects. A crisis in one currency can trigger instantaneous, correlated movements across seemingly unrelated currencies as global investors flee to perceived safe havens. Contagion spreads on the basis of risk sentiment, not economic fundamentals.

International bond markets entangle nations through shared investor bases. When global investors reduce exposure to emerging market debt, they often sell across multiple countries simultaneously. This creates correlated selling pressure that impacts countries with vastly different economic conditions, punishing all simply for being categorized similarly.

Finally, the operational machinery of global trade, such as letters of credit, guarantees, export financing, and so on, creates a web of obligations that connect institutions across every supply chain. A disruption in one hub does not create a ripple effect; instead, it immediately causes stress throughout the entire system, halting commerce before it stops the physical movement of goods. The financial system measures the problem, and the entire entangled network collapses into a state of crisis at once.

The domino effect at light speed: how panic spreads

Financial contagion is the ultimate expression of quantum-like entanglement in action. It demonstrates that problems don't just spread through the system; they instantly manifest across it. It moves at a speed and scale that traditional, linear cause-and-effect models cannot explain. To understand contagion is to accept that financial systems exhibit nonlocal correlations; an event in one location can immediately alter conditions everywhere, defying classical logic and overwhelming conventional defenses.

Crisis transmission mechanisms

Contagion doesn't travel through a single channel. Instead, it attacks through multiple pathways at once. This creates a quantum-like effect where the systemic response is far greater than the sum of its individual transmission paths. Traditional finance understands direct financial linkages well, but quantum finance reveals that the indirect, psychological effects are often more dangerous. This is simply because they are invisible, unpredictable, and impossible to contain.

Credit markets offer a clear example of direct contagion. Losses in one sector, for instance, in commercial real estate lending at Quantaland Bank, cause an immediate tightening of lending standards for all borrowers, not just those in the affected sector. The issue intertwines with the economy's overall credit availability.

Funding markets reveal a similar pattern. During a crisis, these markets can freeze across multiple institutions simultaneously, despite significant differences in their individual credit quality. The fates of these markets become entangled due to their shared dependence on a common resource that has suddenly vanished.

Most potent is information contagion. News of trouble at one investment bank instantly raises suspicions about all others, regardless of their specific exposures. This is the essence of entanglement: a correlation that completely ignores fundamental business relationships, causing panic to spread rapidly.

Market psychology and panic spreading

Psychological contagion creates near-pure quantum effects, correlating institutions through shared fear rather than shared fundamentals. Emotion spreads through markets faster than fact, entangling entities that have no direct business connections. Social media acts as a massive accelerator for this phenomenon, allowing rumors and panic to go viral globally within hours, forcing market reactions long before institutions can respond with factual information.

Algorithmic trading serves as an automated panic amplifier. These systems execute preprogrammed responses to market stress signals, simultaneously acting to reduce risky exposures across multiple assets and markets. This creates a correlated selling pressure that mimics fundamental entanglement but is actually driven by technological correlation caused by machines reading other machines.

This culminates in herd behavior, where investors imitate others' actions without independent analysis. The rational individual decision to avoid losses produces a collectively irrational outcome. The modern bank run illustrates this perfectly. A viral rumor about Quantaland Bank sparks a panic. Depositors, fearing the same could happen elsewhere, withdraw funds from ATMs. The machines run dry, fueling the rumor further and creating long queues before the bank even opens. The teller counters are swiftly drained. What started as a perceived risk has turned into a genuine liquidity crisis due to the actions of the depositors themselves. Psychological entanglement has overridden economic reality.

Regulatory responses and their limitations

Regulatory attempts to contain contagion frequently fail because they are designed for a classical world of isolated events, not a quantum-entangled network. They presume that we can wall off problems, but entanglement effects simply navigate these barriers through unexpected channels. During a crisis, traditional regulatory tools are often too slow to even identify the problem before it has already spread.

Consider deposit insurance. While it successfully prevents runs at insured institutions, it can inadvertently reshape entanglement rather than eliminate it. It encourages depositors to flee from uninsured institutions to insured ones, simply transferring the pressure and creating a new systemic fault line. The solution changes the nature of the problem but does not resolve it.

Cross-border regulation faces an even greater challenge. National authorities must respond to a globally entangled system. An action taken in one jurisdiction, for example, a tariff hike by Quantaland's largest trading partner, can trigger unintended consequences in others, as politically independent nations adjust their policies in response. The speed of financial entanglement vastly exceeds the sluggish pace of international regulatory cooperation.

Speed of contagion in the digital age

Digital technology has supercharged the velocity of contagion, creating an environment where problems propagate globally faster than human decision-makers can even perceive them. Electronic trading, automation, and global connectivity have compressed the time frame for systemic collapse from weeks to minutes.

High-frequency trading algorithms can detect a stress signal and liquidate billions in assets within milliseconds. Technological herd behavior drives correlated market drops that look like fundamental entanglement.

The operational infrastructure itself is a source of entanglement. The global payment systems are so interdependent that a disruption in one major network instantly cripples supply chains and financial relationships worldwide, ahead of implementing any contingency plan.

Finally, cyber threats represent the newest and most dangerous vector. A security breach at Quantaland Bank, for example, doesn't just create a local IT problem for one institution. It triggers defensive responses across the entire financial ecosystem, disrupting global counterparties and customers instantly. In the digital age, a single point of failure can entangle the entire world.

The architecture of collapse: network theory in finance

Financial decisions create quantum-like entanglement through vast networks of obligations, shared exposures, and interdependent strategies. To understand the resultant cascading effects, you must view the financial system through the lens of network theory. You need to accept that individual actions can trigger system-wide responses that far exceed the expected magnitude of the initial event.

Network theory in finance: too connected to fail

Financial networks exhibit fundamental quantum-like properties. The connections between institutions influence the system's behavior more than their individual traits. Network theory provides the essential tools for mapping these connections and understanding how they create systemic vulnerabilities. However, these models themselves struggle to capture the dynamic, evolving nature of quantum financial relationships.

Mapping financial networks

Mapping these networks reveals the hidden architecture of entanglement, connecting institutions through both direct ties (lending, derivatives, payments) and indirect ones (shared exposures, common counterparties, correlated strategies). Crucially, these relationships are not static. During stable periods, institutions maintain loose coupling that allows for independent operation. Under stress, these connections tighten instantly, creating strong correlations that characterize quantum entanglement.

An analysis of Quantaland's financial network shows how apparent diversification can conceal dangerous concentrations. When Quantaland Bank lends to several different companies that all depend on the same critical supplier, it creates indirect entanglement. Trouble at that single supplier can now trigger correlated defaults across multiple borrowers, reflecting a quantum-like correlation that traditional, isolated credit analysis would completely miss.

Derivative networks create particularly complex entanglement. A single trade can create obligations linking dozens of institutions through clearinghouses, prime brokers, and collateral chains. These connections typically remain invisible, hidden in the lining of the system, until a major default forces the network to reveal its true, entangled structure.

Hub institutions and systemic importance

Network analysis identifies hub institutions whose central position gives them a disproportionate impact on the entire system. These hubs function like quantum measurement devices; their distress instantly transforms the network into a stressed state.

While "too-big-to-fail" banks are obvious hubs, connectivity is often more important than size. Institutions that serve as critical intermediaries or clearinghouses can create massive systemic vulnerabilities regardless of their balance sheet. Their network position, not their size, makes them systemically important.

Regulation can even create artificial hubs. Central counterparties (CCPs) were designed to concentrate and manage risk in supervised entities. While this aims to improve stability, it creates a new form of entanglement, making every network participant dependent on the continued health of a single, massive hub.

Prime brokerage relationships explain this hub effect. If a prime broker in Quantaland suffers a significant loss, it issues margin calls to all its hedge fund clients simultaneously. This call triggers forced selling across unrelated strategies, driving down asset values and creating correlated stress through shared dependence on the hub, not through fundamental links.

Central banks and major correspondent banks are the ultimate operational hubs. Their failure would cause a quantum-like collapse where entire domestic payment or international transaction networks become inoperative at once. A simple glitch at Quantaland Central Bank can halt national transactions, creating instantaneous, economy-wide paralysis.

Network fragility and resilience

Financial networks possess quantum-like properties where stability can vanish in an instant. Fragility emerges from the pattern of interconnections themselves, not necessarily from the weakness of individual nodes. This makes systemic stability almost impossible to assess with traditional analysis.

Cascading failure is the major risk. An initial loss at one institution propagates through its network of obligations, threatening the stability of its counterparties, who then threaten theirs. This entanglement can amplify a local failure into a full-blown systemic crisis.

A network's resilience depends on the diversity of its connections and the availability of alternative pathways. Robust networks have multiple, varied connection types that are unlikely to fail simultaneously. Fragile networks rely on similar connections that can fail in unison under stress, correlating the entire system's collapse.

These networks exhibit critical threshold effects. Below a certain stress level, they can absorb shocks. Beyond that level, they undergo a quantum-like phase transition. This causes a sudden shifting from a stable to an unstable state and experiencing cascading failures that traditional, linear models cannot predict.

Limitations of network models

Despite their power, network models face fundamental challenges when applied to finance's dynamic reality. Their greatest weakness is static analysis. Models capture a snapshot of connections at a specific time. But financial relationships are alive and dynamic. They strengthen, weaken, and evolve with market conditions, rendering network maps obsolete quickly, especially during crises.

Incomplete information is a crippling constraint. Critical relationships such as derivative exposures, off-balance-sheet arrangements, contingent obligations, and so on are often private or invisible. Network models that miss these hidden links can develop a fatal false optimism about the system's stability.

Behavioral assumptions, too, are often flawed. Models assume institutions will respond to stress in predictable ways. They fail to account for how quantum entanglement creates unexpected behavioral correlations. When every institution simultaneously abandons normal practices for defensive survival, it creates new, self-reinforcing patterns of crisis.

Finally, these models create observer effects. When the Federal Reserve of Quantaland designates Quantaland Bank as systemically important based on network analysis, it changes how every other institution interacts with it. The act of measurement itself reshapes the network being measured, creating a quantum feedback loop that undermines the model's predictive power.

The hidden web: counterparty risk as an engine of entanglement

Counterparty risk is the engine of financial entanglement. It forges correlations between institutions through chains of mutual obligations, even when no direct business relationship exists. To manage it, you must recognize that these chains can create system-wide vulnerabilities that far exceed the apparent risk of any single transaction.

Direct and indirect exposures

Direct exposures such as loans, bonds, derivatives, and so on create obvious entanglement, where one party faces immediate loss if another fails. But the true danger lies in indirect exposures. They are harder to identify and manage, and they are often far more systemic.

Consider Quantaland Trust Company. They meticulously manage direct exposure to each counterparty with credit limits and collateral. However, their indirect exposure through investment funds creates hidden entanglement. If multiple funds in their portfolio hold the same stock for diversification, they have inadvertently created a concentrated, correlated exposure. Diversification designed to reduce risk has instead amplified it through a network effect.

Chain exposures create even deeper indirect links. If Quantaland Investments is exposed to Quantaland Bank, and Quantaland Bank is exposed to Quantaland Manufacturing, then Quantaland Investments is now entangled with Quantaland Manufacturing's fate. These chains can extend across dozens of institutions, meaning problems in a distant, unrelated company can ripple back to you. This is how, in 2008, subprime mortgage problems devastated institutions with no direct mortgage exposure.

Contingent exposures play their role too. Credit lines, guarantees, and certain derivatives may show no current exposure but can generate sudden, massive obligations the moment a triggering event occurs. These relationships lie dormant until a crisis activates them simultaneously across the entire system.

Collateral chains and rehypothecation

Rehypothecation, which involves reusing collateral from one transaction to secure another, generates a complex network of quantum entanglement. It allows the system to operate with less capital, but it forges links where a problem with one asset can instantly jeopardize multiple institutions.

At Quantaland Securities, hedge funds post collateral for margin. The firm can then rehypothecate that same collateral to meet its obligations to clearinghouses. This entangles those hedge funds with the risks of Quantaland Securities' counterparties, creating a hidden web of dependency.

The same principle applies in lending. Quantaland Bank lends to Quantaland Manufacturing against collateral and then refinances by pledging that same collateral to Quantaland Investments. It has now directly entangled Quantaland Investments with the manufacturer's financial health. There is no direct link, but their fates are now correlated.

Collateral Asset Management, by converting lower-quality assets into acceptable collateral, creates an entire ecosystem of dependency. Imagine a U.S. bank needing Quantaland bonds. The U.S. bank provides a

Quantaland Bank with $120 million in U.S. corporate bonds in exchange for $100 million in Quantaland bonds. The Quantaland Bank then repos the U.S. bonds to fund the position. If those U.S. bonds fall in value, the U.S. bank gets a margin call. The Quantaland Bank suffers losses on its repo book and may struggle to fund the position, triggering adverse reactions in the repo market.

A problem in one asset class instantly propagates through the chain, entangling institutions across borders and markets.

Central clearing and systemic concentration

Central CCPs were created as a regulatory solution to bilateral counterparty risk. But by concentrating obligations, they end up creating a new form of entanglement, making every participant dependent on the health and judgment of a single, massive hub.

CCPs interpose themselves in every trade, becoming the buyer to every seller and the seller to every buyer. This eliminates bilateral risk but creates a monumental point of failure. The collapse of a major CCP would trigger an instantaneous, system-wide catastrophe.

Their sophisticated risk-management systems, including real-time exposure monitoring and dynamic margining, create their own form of entanglement. Every participant is now dependent on the CCP's models being correct and its default reserves being adequate. Most dangerously, CCP margin systems are pro-cyclical. In a crisis, volatility triggers margin calls for all participants simultaneously. This forces widespread liquidations, which increase volatility further and trigger higher margin calls. The system designed to control risk instead amplifies it through a destructive feedback loop.

Ultimately, the "default waterfall" causes all other members to share in the losses resulting from one member's failure. CCPs transform idiosyncratic risk into systematic risk. The failure of one becomes a problem for everyone.

Credit default swaps and risk amplification

Credit default swaps (CDS) represent the ultimate paradox: an instrument designed to distribute risk instead often amplifies and concentrates it, creating complex entanglement patterns.

A CDS entangles the protection buyer, the seller, and the reference entity. A problem at the reference entity doesn't just cause a loss for the debt holder; it triggers obligations for the protection seller, correlating their fates.

This entanglement becomes truly systemic when institutions sell CDS protection speculatively, without owning the underlying debt. They become pure speculators on credit risk, weaving a web of obligations that connects them to entities they otherwise have no link to. When a major default occurs, it doesn't just burn the debt holders; it triggers widespread losses across this vast, interconnected web of speculators, amplifying the initial shock into a system-wide event.

Meanwhile, far from Quantaland

Evergrande crisis (2021)

The Quantum State: Evergrande's debt problems created a state of financial entanglement where the Chinese real estate, global commodity markets, and international banking sectors became correlated. This correlation transmitted stress across seemingly disconnected systems.

The Story: China Evergrande Group's debt crisis in 2021 revealed the complex web of entangled financial relationships connecting Chinese real estate to the global economy. As China's second-largest property developer, Evergrande had amassed approximately $300 billion in liabilities by mid-2021. This was equivalent to roughly 1.7% of China's GDP. This massive debt load was distributed allegedly across 128 banks and 121 non-banking institutions. This was an intricate network of financial exposure.

In September 2021, Evergrande warned investors of cash-flow issues and potential default risks. This announcement triggered immediate contagion effects across global markets. Within days, Hong Kong's Hang Seng Index dropped more than 3%, the S&P 500 fell 1.7%, and European bank stocks declined around 3–5% due to their Chinese exposure. Interestingly, these markets did not have any direct exposure to Evergrande's debt. The markets reacted to broader systemic risk, far beyond Evergrande alone.

The entanglement extended far beyond financial markets. Chinese construction activities were expected to slow down, which led to iron ore prices going down by nearly 20% in September 2021. Australia's mining-heavy stock index fell more than 2%, while Brazilian mining stocks saw share price declines of around 5%. Evergrande's problems affected commodity producers on different continents. This interconnectedness was particularly evident in the divergent market reactions based on varying assessments of potential contagion paths.

Chinese authorities ultimately intervened to manage the crisis. They encouraged state-owned enterprises to acquire Evergrande assets and directed banks to maintain credit lines to the property sector. This intervention aimed to contain the contagion by breaking the most dangerous entangled relationships. By early 2022, the immediate global panic had subsided; however, Evergrande was still engaged in restructuring negotiations with its creditors. The property developer officially defaulted on its dollar bonds in December 2021. This contributed to a slowdown in China's real estate sector and went on to reverberate through the economy.

The Quantum Finance Lesson: The Evergrande crisis provides a vivid example of how financial entanglement can swiftly transmit shocks through global markets, regardless of geographical boundaries. Much like quantum entanglement, where particles influence each other across vast distances, disruptions within one part of the financial system can ripple outward in a "nonlocal" manner. This mirrors the quantum principle of nonlocality, in which entangled particles affect one another no matter how far apart they are. Complex layers of intermediaries in the financial world reveal this nonlocality. Layers such as counterparty exposures, cross-border investment flows, lengthy supply chains, and others act as channels for contagion.

In these quantum-entangled financial systems, shocks and price movements can cascade with remarkable speed, giving rise to the sense of "instantaneous action at a distance." As a result, turmoil in one sector can almost immediately influence assets and markets across the globe that might otherwise seem unrelated.

Practical Application: As a finance professional, you should incorporate quantum thinking by mapping entangled relationships between assets rather than analyzing them in isolation. You should identify how unrelated investments might become correlated during stress events and construct portfolios that account for these hidden connections.

This quantum perspective acknowledges that financial markets, like quantum systems, demonstrate nonlocal correlations that defy classical risk models. You may ignore this exclusively at your own risk.

2008 financial crisis

The Quantum State: The U.S. subprime mortgage market became entangled with the global financial system through a complex securitization process. This created a state where defaults in American suburbs instantaneously affected banks worldwide.

The Story: The 2008 financial crisis represents perhaps the most profound example of financial entanglement in modern history. What began as a correction in the U.S. housing market escalated into a global financial catastrophe that nearly collapsed the international banking system. The profound and complex interconnections within the modern financial system were evident to everyone.

The crisis originated in the U.S. subprime mortgage market. In the early 2000s housing boom, lenders extended loans to borrowers with questionable creditworthiness. These risky mortgages were then packaged into complex securities known as mortgage-backed securities (MBS) and collateralized debt obligations (CDOs). These instruments were sold to investors globally, creating a backdoor to global contagion. The innovation of CDS created additional layers of entanglement. CDOs allowed institutions to trade the risk of default without owning the underlying securities. Risk became a commodity.

As U.S. housing prices began declining in 2006–2007, mortgage defaults increased. This was more prevalent among subprime borrowers. By 2008, what might have remained a contained problem in one sector of the U.S. economy had become a global financial pandemic. The entanglement became dramatically visible when Lehman Brothers filed for bankruptcy on September 15, 2008, triggering immediate worldwide effects.

Stock markets worldwide plummeted, with the Dow Jones Industrial Average falling 4.4% on September 15 alone. Within a month, the MSCI World Index had lost nearly 20% of its value. Banks globally faced liquidity crises regardless of their direct exposure to U.S. mortgages. The Treasury-Eurodollar spread spread, a key measure of credit risk, soared from 1.1% in September to 4.6% by October, reflecting the trust deficit in interbank lending.

Central banks were forced to coordinate unprecedented actions. To address the shortage of USD at a global scale, the Federal Reserve established currency swap lines with the European Central Bank (ECB), the Bank of England, and other central banks.

The crisis revealed how financial innovation had created hidden connections across institutions, sectors, and countries. Perhaps most telling was how defaults on mortgages in places like Nevada and Florida could threaten banks in Germany and Japan.

The aftermath led to global regulatory reforms, most notably the Basel III accords and the Dodd-Frank Act in the United States. These regulations specifically addressed systemic risk and sought to reduce dangerous forms of financial entanglement. By this time, you know that regulations actually set up another wave in motion.

The Lesson of Quantum Finance: The crisis of 2008 stands as a defining example of financial entanglement in action. Complex interconnections across global finance allowed shocks to ripple instantly between institutions and markets that, on the surface, appeared unrelated. Innovations in finance had developed channels where a mortgage default in an American suburb could immediately impact banks on other continents. This phenomenon closely resembles quantum entanglement, where a change to one particle instantly influences its entangled partner.

The crisis also illustrates the concept of entanglement swapping from quantum theory: particles that have never met can become connected through intermediaries. In financial terms, this is analogous to how layers of securitization and derivatives form a web that allows mortgage borrowers and global investors without any direct link to be bound together by complex financial instruments.

The events of 2008 made clear that entanglement is not a rare occurrence but a central feature of the world's modern financial systems.

Practical Application: Financial regulators and risk managers should incorporate quantum thinking by focusing on entangled relationships rather than isolated risks. They need to map the complex web of direct and indirect exposures that connect institutions. They should further implement stress tests that account for these nonlinear relationships.

The quantum perspective acknowledges that financial stability depends not just on the health of individual institutions but also on the nature of their entanglements and how the network behaves.

European debt crisis (2010–2012)

The Quantum State: European sovereign debt markets became entangled through monetary union and banking system interconnections. This created a state where fiscal problems in Greece instantaneously affected borrowing costs across the Eurozone.

The Story: The European sovereign debt crisis that unfolded between 2010 and 2012 provides an illustration of financial entanglement within a monetary union. The crisis began as a fiscal problem in Greece, which represented just 2% of the Eurozone's GDP. It eventually escalated into an existential threat to the euro itself, demonstrating how deeply interconnected the fates of the union's member states are.

In late 2009, the newly elected Greek government revealed that previous administrations had misreported budget data, triggering the crisis. The Greek deficit for 2009 was not 3.7% of GDP as previously reported but was around 15.4%. This was more than five times the Eurozone's limit. This discovery sparked immediate contagion across European bond markets as investors reassessed sovereign risk.

Entanglement became increasingly visible throughout 2010. Greek bond yields soared from approximately 5% in December 2009 to over 12% by May 2010, despite the announcement of a €110 billion bailout package from the EU and IMF. Simultaneously, yields on Portuguese, Irish, Spanish, and Italian bonds rose sharply, regardless of their differing fiscal situations.

This market entanglement reflected several underlying connections. European banks held significant amounts of sovereign debt from multiple Eurozone countries, creating direct transmission channels for contagion. French and German banks alone had around €100 billion in exposure to Greek debt. The shared currency also meant that individual countries couldn't use monetary policy or currency devaluation to address fiscal problems. This made defaults more likely across vulnerable members. In addition, market psychology created self-reinforcing feedback loops as investors began viewing the asset as a single asset class rather than as distinct economies.

The crisis reached its peak in 2012 when concerns about a potential breakup of the euro sent Spanish 10-year bond yields above 7% and Italian yields above 6%. The debt burdens were unsustainable at this level. The turning point came in July 2012 when the ECB announced the Outright Monetary Transactions (OMT) program. This effectively broke the most dangerous aspects of market entanglement by providing an implicit ECB backstop to sovereign debt markets. Bond yields quickly declined across the periphery. The spread between German and Spanish 10-year bonds contracted from over 600 basis points in July 2012 to around 350 basis points by year-end.

The subsequent years saw continued efforts to reduce harmful forms of entanglement while maintaining beneficial integration. These included the establishment of the European Stability Mechanism, the launch of the Banking Union, and enhanced fiscal coordination.

The Quantum Finance Lesson: The European debt crisis powerfully demonstrates the phenomenon of financial entanglement within a currency union. It highlighted how shared monetary policy and closely linked banking systems can create quantum-like relationships among fiscal entities. Much like entangled particles in quantum physics, which remain interconnected regardless of their distance, countries within the Eurozone saw their borrowing costs bound together. It was immaterial that their underlying economies differed greatly.

This is analogous to the quantum concept of entanglement, where observing one particle instantly affects its partner. In the Eurozone, when markets reevaluated Greek debt, the borrowing rates for nations like Spain and Italy shifted immediately, transmitted through a complex network of financial and institutional ties. The single currency made these economies operate as an entangled system where risk was perceived collectively, not individually.

Practical Application: Policymakers should incorporate quantum thinking by designing financial architecture that harnesses beneficial forms of entanglement. At the same time, they should limit destructive correlations. This means creating firewalls between interconnected systems that can be activated during crises while maintaining integration benefits during normal times. Many such mechanisms are already in place, suggesting quantum-like thinking.

This quantum perspective acknowledges that financial systems, like quantum systems, can exhibit both beneficial and harmful forms of entanglement. What is even more important is to realize that these entanglements cannot be eliminated.

Looking ahead

In this chapter, we saw how quantum entanglement makes seemingly independent financial institutions dance to the same destructive rhythm. The stories ranged from Florida mortgages threatening German banks to Greek debt shaking Italian bonds to Chinese real estate moving Australian mining stocks. This web of hidden connections transforms diversification from protection into illusion, revealing that in our interconnected world, your risk truly becomes everyone's risk. But this entanglement creates an even deeper challenge for financial professionals: if the system is this interconnected and unpredictable, how can we possibly plan for the long term? In our next chapter, we'll confront the uncomfortable truth that long-term stability itself may be the greatest illusion of all. You'll discover why Kodak's 130-year photography dominance collapsed in a decade, how Netflix's quantum approach defeated Blockbuster's rigid planning, and why the very pursuit of certainty through detailed long-term forecasts may be setting you up for catastrophic failure. Get ready to challenge all your previous beliefs about strategic planning and risk management.

Chapter 7

Risk, uncertainty, and the fallacy of long-term stability

"*The sailor checks the clouds from the port and the waves from the deck*," goes an old Quantaland proverb.

Entrepreneur Aisha built her startup fashion-designing house, carefully studying fashion trends over the last 50 years. She never thought that the business model would crumble. It was impossible to believe that e-commerce platforms would take over a business that has always been a bespoke service. Past trends betrayed her. This chapter reveals why long-term stability is an illusion. You will review quantum analogies and cases like Kodak and Blockbuster to see how businesses fall by treating uncertain futures as certain. You will be exposed to hidden biases in data and find how quantum-inspired frameworks for strategic flexibility are a preferred option over a long-term plan.

The mirage of certainty: why long-term forecasts fail

Long-term financial planning operates under a fundamental quantum paradox: the precision of your prediction is inversely proportional to the length of your time horizon. The further out you look, your sophisticated, detailed models become exercises in false precision, creating dangerous illusions of control over an inherently uncertain future.

Quantum finance does not ask you to abandon long-term planning. It demands that you reinvent it. This approach requires you to reject the assumption of a single, predictable future based on extrapolating the present. Instead, you must build decisions around flexibility and optionality. Your models should not attempt to forecast the outcome but acknowledge and prepare for a spectrum of possible future scenarios.

The tyranny of the past: when data become a trap

Financial analysis suffers from a fundamental quantum measurement problem. The very act of observing historical data changes both our

DOI: 10.4324/9781003619635-7

understanding of the markets and the markets themselves. This creates systematic biases that render the past a less reliable guide to the future. Traditional finance often overlooks this and treats the measurement of the past as an infallible blueprint for what is to come.

The illusion of predictive patterns

Historical data create a dangerous illusion of predictability. Past patterns appear to be solid forecasts, but they are actually a quantum-like mix of possibilities. They simultaneously represent genuine relationships and mere statistical coincidences. Only future events will reveal which one they were.

Consider Quantaland Pension Fund. It built a bold, equity-heavy strategy based on 50 years of data showing stocks beating bonds in 42 of those years. But this historical record was just one realized path, a single quantum outcome that collapsed a spectrum of possibilities into one sequence. The management became blind to other potential realities, like a prolonged bond rally. Markets don't replay the past; they are a web of possibilities waiting for an event like a crisis or an innovation to lock in a new reality.

This observer effect is clear in technical analysis. When Quantaland Traders all recognize the same "support level" on a stock chart and buy accordingly, their collective action pushes the price up. This ends up breaking the very pattern they bet on. The observation altered the support level and subsequently affected the trading design.

The problem intensifies with algorithmic trading. Strategies based on historical indicators, like the 200-day moving average used by Quantaland Algo Fund, become self-defeating through widespread adoption. If every algorithm follows the same trend, their correlated buying triggers a frenzy that inevitably culminates in a sudden, violent breakdown of the pattern. Markets don't transition gradually between regimes; they leap from one state to another.

Even asset correlations are not stable laws. The relationship between technology stocks and bonds, seemingly stable for years, can flip instantly during a crisis. This occurs because both technology stocks and bonds become entangled due to shared investor panic. The diversification you relied on vanishes, revealing a correlation that never appeared in historical data.

Survivorship bias in financial analysis

Financial databases present a curated history, systematically excluding failed companies, discontinued funds, and abandoned strategies. This is a form of quantum measurement error where we study only the survivors, profoundly distorting our perception of risk and opportunity. We learn from triumphs and ignore the failures.

Index performance is a classic example. The constant removal of failing companies and addition of winners make the index's long-term growth appear smooth and inevitable. In the bargain, it hides the constant churn and failure inherent in a dynamic economy.

Quantaland Asset Management made this error when reviewing its 20-year track record. Excluding all the strategies and funds abandoned due to poor performance inflates the perception of their skill. It is like trying to understand airplane crashes by only studying planes that landed safely.

The same bias plagues mutual fund and hedge fund databases. The silent removal of the worst performers from the sample skews the "average" return upward. Venture capital is perhaps the most extreme case, where a few highly visible winners like Quantaland Startups create the illusion that startup investing is predictable. The reality is that the vast majority of failures disappear without a trace.

The quantum nature of financial time series

Financial time series behave less like classical mechanical systems and more like quantum systems, where multiple potential future paths coexist until a market event collapses them into a single reality. This explains why traditional statistical analysis so often fails when we need it to work the most.

There are various market realities that evidence this. High-volatility periods cluster together through feedback loops. This creates "fat-tailed" distributions where extreme events occur far more frequently than Gaussian models predict. Similarly, we expect and even witness today's returns influencing tomorrow's. But these correlations can vanish instantly when market conditions shift, causing mean-reversion strategies to collapse without warning.

Like quantum entanglement, distant events remain connected. A crisis from a decade ago can still influence market psychology and liquidity today through hidden channels that simple historical analysis misses entirely. Furthermore, the fundamental parameters of the market, such as volatility, correlation, drift, and so on, remain dynamic. They shift without warning, rendering models built on past "steady states" useless precisely during the stress periods they were designed to navigate. Linearity is convenient to assume but is not the only truth.

Quantaland Trading learned this harshly. Their momentum algorithms profited for five years based on historical patterns. When the market conditions experienced a significant shift into a new regime, those same signals resulted in overnight, massive losses. There was no gradual transition, only a sudden, decisive break with the past.

The unknowable unknown: in search of black swans

Black Swan events are the ultimate quantum-like phenomena in finance. They are extremely rare occurrences that defy prediction and have a catastrophic, system-wide impact. They represent a fundamental failure of traditional risk management because they exist outside all normal probability distributions and historical analysis.

Defining true unpredictability versus low probability

A Black Swan is not a low-probability event; it is fundamentally an unpredictable one. You can model rare events, like a hurricane, using historical data to estimate frequency and severity. But a Black Swan is a quantum-like surprise that exists outside our conceptual frameworks, becoming visible only in its devastating aftermath.

Quantaland Insurance learned this distinction brutally. Their catastrophe models could price hurricane risk using centuries of weather data. However, COVID-19 was a truly exceptional occurrence. This was a pandemic with no modern precedent that forced global economic lockdowns. Their models were blind to it, as there was no availability of a reference class. Hurricanes are a known category of disaster. A pandemic that halts global commerce represented a quantum-like transition to an entirely new economic state, for which no data existed.

Market crashes operate the same way. The 1987 Black Monday, far from being a slide, was a sudden, discontinuous leap driven by program trading and portfolio insurance. These factors were invisible until they triggered the collapse. Similarly, technological disruptions like the internet didn't just improve commerce; they created a whole new reality that was impossible to predict from the prior paradigm.

True unpredictability emerges from the quantum-like nature of complex systems, where tiny, unseen shifts can trigger massive, nonlinear reactions. Black Swans don't stretch historical models. They actually shatter them, revealing realities those models could never conceive.

The impact asymmetry of extreme events

Black Swans exhibit a profound quantum-like asymmetry. The pace and intensity of devastation significantly surpass the gradual process of creation. Adverse outcomes can wipe out decades of gains almost instantly, a lopsided effect caused by financial entanglement. This entanglement permits fear and failure to propagate through networks at light speed.

Traditional portfolio theory assumes a symmetric world where gains and losses mirror each other. The reality is quantum-like, where losses can be bottomless, while gains are inherently capped. Quantaland Venture Capital lived through this. A winning investment might take years to deliver a tenfold return, but a failure incinerates 100% of the capital in weeks. Portfolios can collapse far faster than their construction due to this timing mismatch.

Leverage acts as an amplifier on this asymmetry, creating quantum thresholds where a small dip can trigger a total wipeout. Strategies that thrive in calm markets disintegrate when volatility defies all historical norms.

This asymmetry also defines recovery. Systems collapse in days or weeks but take years to rebuild. Confidence vanishes overnight and returns only grudgingly. Markets don't simply bounce back; they undergo a long, painful descent into a new, often diminished, state.

Why normal distributions fail in finance

Financial returns behave nothing like the smooth, predictable bell curves of normal distributions. They are quantum systems, prone to sudden, discontinuous leaps. The most obvious failures are fat tails. Extreme events with high orders of magnitude occur more frequently than Gaussian models predict. These are not merely significant fluctuations; they represent quantum leaps into entirely new market regimes, making the past obsolete.

Quantaland Bank's VaR models were a perfect victim. Based on normal distributions, they calculated that a Q$10 million daily loss should occur roughly once a year. In reality, during periods of stress, such losses occurred every few months. The model, beyond just being wrong, was dangerously misleading.

Volatility clustering creates a quantum-like feedback loop, where large swings beget more large swings, violating the core assumption of independent price movements. Add to that another critical flaw, skewness. Return distributions are not symmetrical; large drops happen more frequently than large gains. This systematically fools risk models into underestimating downside risk.

Finally, these distributions exhibit jump risk. There are price gaps that resemble quantum leaps where assets reprice instantly to new levels based on a shock. It skips any gradual transition. The market, instead of always moving gradually, often teleports into a new and unknown future.

The perils of prediction: how certainty creates vulnerability

Long-term financial planning forces you to make assumptions about the future that eventually determine the success or failure of your current decisions. Quantum finance reveals that these assumptions don't represent a single future. In fact, they exist in the superposition of multiple potential realities, waiting for unforeseen events to collapse them into one. Relying on static predictions treats these quantum possibilities as fixed certainties, creating massive vulnerability when the actual future inevitably diverges from your single, linear projection.

Interest-rate assumptions

There are few better examples where this quantum uncertainty is more apparent than in forecasting interest rates. The challenge of predicting them creates unavoidable exposure for any institution deciding about the tenure of its assets and liabilities. Recognizing this fundamental uncertainty is the first step to managing the immense risk that traditional static approaches so often miss.

Asset-liability mismatches and duration risk

Duration risk is often treated as a puzzle to be solved with better forecasts. In reality, it is a fundamental quantum uncertainty in future rates that you must navigate. Institutions end up locking in today's rates to fund tomorrow's obligations, betting against an unknowable future. This is not a puzzle to be solved but a fundamental risk to be managed with any duration-based position.

Quantaland Pension Fund faces this classic problem. Their actuarial models precisely calculate that they need a 7.2% annual return to meet obligations. Unfortunately, this precision in target setting creates dangerous inflexibility in investment strategy. The fund's commitment to specific asset allocations between stocks and bonds reflects their actuarial models' bet on steady long-term returns. But interest rates don't follow a predictable path. They hover in a superposition of multiple possibilities until central bank policy or an economic shock forces a specific outcome.

This creates a quantum asymmetry. When rates rise, the value of long-duration assets plummets more than the value of long-duration liabilities, creating immediate net losses. When rates fall, the gains on assets may not fully offset the skyrocketing present value of future liabilities.

The problem intensifies with structural mismatches. Quantaland Insurance sells life policies, collecting premiums today to pay claims years later. This inevitably creates duration risk. Rising rates may boost their investment income but not enough to cover the increased cost of maintaining policy reserves. Falling rates create the opposite trap. Pension obligations exhibit a pure quantum exposure, where benefit payments remain fixed, yet the assets supporting them yield uncertain returns. Small changes in your discount rate assumption can alter funding requirements dramatically, creating an entangled system where assets and liabilities are locked in a strange dance, with none knowing where the next step will be.

The impossibility of predicting rate cycles

A fundamental quantum uncertainty plagues interest-rate forecasting. People's expectations of rates shape economic conditions, which central banks use to set policy. This creates a feedback loop where the forecast influences the outcome, making long-term predictions inherently unreliable.

Central banks operate in a quantum-like state of superposition, with multiple policy options on the table. Eventually, a new inflation report or jobs number acts as a measurement and collapses those possibilities into a single decision.

Quantaland Bank has been through this uncertainty. While its long-term deposits lock in at fixed rates, its commercial loans are subject to volatile market rates. Predicting the gap between them means guessing the path of the entire economy and the reactions of competitors, which undoubtedly are a near-impossible task.

Economic cycle forecasting aims to predict patterns in rates by analyzing historical data. However, shifts in the quantum-like regime constantly upend the relationships between economic factors. This renders the past useless. Furthermore, domestic rates are entangled with global capital flows and foreign policy decisions. Rates in Quantaland don't exist in isolation but respond to a chaotic international system.

Quantum-like swings can even affect the yield curve, which represents the market's collective expectation for future rates. Overnight, new data can upend the consensus view. Today's curve's forward rates reflect the current expectations, not the unknown future states that upcoming shocks will reveal.

Floating rate complexities and basis risk

Floating rate instruments seem to solve duration risk by adjusting with market rates. However, they merely exchange one type of risk for a more complex form of uncertainty known as basis risk.

These instruments create quantum-like entanglement by tracking reference indices (like Secured Overnight Financing Rate, commonly known as SOFR) that may not perfectly match an institution's actual funding costs. The LIBOR scandal was a perfect example, showing how the supposed benchmark could diverge from reality.

Quantaland Corporate faces this in its floating-rate debt. Its loans reprice off the prime rate, but its funding comes from long-term deposits at fixed rates. Forecasting the spread between them requires predicting unpredictable market and competitive shifts. We have touched upon this earlier too.

Reset frequency mismatches add another layer of timing risk. If assets are repriced monthly but liabilities are adjusted quarterly, short-term gaps can quickly accumulate into significant exposures during periods of volatility.

Finally, embedded options like caps and floors introduce quantum-like nonlinearities. Payments remain unchanged until rates move past a set level, at which point the cash-flow impact can be sudden and severe. Even credit spread volatility acts independently, often widening during crises even as base rates fall. This adds a second, unpredictable dimension of risk that floating rates alone cannot eliminate.

When technology shatters linearity

Technology doesn't follow a predictable path of innovation and creates quantum-like disruptions. Breakthrough innovations can instantly collapse existing business models and render long-term investments obsolete. They behave less like a gradual evolution and more like a sudden phase transition.

The exponential nature of innovation

Technological progress exhibits quantum behavior, advancing exponentially and breaching critical thresholds without warning. Like a particle leaping to a new energy state, it defies linear projection. Moore's Law is the classic example, where semiconductor capabilities doubled at a pace that consistently shattered linear forecasts. This is akin to quantum tunneling, where technology suddenly bypasses what were once considered impregnable barriers.

Quantaland Technology learned this lesson with its data storage investments. Its 10-year plan, built on assumptions of steady, linear growth, was obliterated in just a few years by an exponential surge in video and Internet of Things (IoT) data. Linear models are useless here and end up creating fatal blind spots in strategic planning.

Network effects act as a multiplier, creating quantum bursts of adoption once the user base hits a critical number. AI is a prime example. For decades, progress was incremental. Then, once processing power crossed an affordability threshold, capabilities exploded in a leap that no one predicted even closely.

The ultimate disruption comes from technology convergence. When separate trends such as GPS, mobile tech, and mapping software fused, they didn't just add together linearly. These trends multiplied into an entirely new reality, leading to the emergence of ride-hiring and ride-sharing solutions. This created an industry that dismantled traditional transportation almost overnight. For a finance professional, the lesson is clear, and you should not risk missing it. Technological disruption doesn't arrive gradually. Technological disruption strikes abruptly, requiring a thorough reevaluation of value and strategy.

Stranded assets and obsolescence risk

Long-term capital investments carry a profound quantum obsolescence risk. Technology can quickly turn a valuable asset into a stranded liability, showing no signs of the usual, predictable, gradual depreciation.

The energy sector is also a positive example. Quantaland Power's coal plants were profitable when built. But as the cost of solar and wind power plummeted exponentially, these assets underwent a quantum-like shift. They became uneconomical long before what their accounting depreciation schedules suggested and the CAPEX evaluation considered.

This pattern repeats across industries. Quantaland Communications invested heavily in fiber-optic networks, only to see low-orbit satellite technology achieve viability and slash demand for fixed-line broadband almost overnight.

Retail real estate faced a similar quantum collapse. Instead of gradually increasing its market share, e-commerce reached its peak quickly. This transformed profitable shopping centers into stranded assets faster than lease terms could adjust. This is a recurring pattern visible over the years. The same retail spaces that once came up, replacing movie theaters made obsolete by streaming services, are now struggling.

From autonomous vehicles threatening parking infrastructure to automation rendering manual assembly lines inefficient, the threat is universal. These quantum shifts make traditional depreciation schedules and capital budgeting models dangerously obsolete. The risk isn't of gradual impairment but of instantaneous irrelevance.

Network effects and winner-take-all dynamics

Technology markets are defined by quantum-like winner-take-all dynamics. Small initial advantages trigger network effects that lead to exponential returns and market dominance. This results in creating astonishing valuations that defy traditional linear models.

Platform businesses such as social media, payment systems, digital marketplaces, and similar ones, with each additional user joining, gain value not linearly but exponentially. This allows them to command massive valuations based on future potential, often long before they turn a profit.

Quantaland Software's collaborative platform faced entrenched competitors. But by achieving rapid, early customer acquisition, it hit a critical mass that triggered a network effect. Its adoption exploded, seizing market share at a pace that traditional competitive analysis could never have forecast.

Superior technology doesn't always win. Standard-setting battles are often decided by quantum uncertainty. Network effects can build unstoppable momentum for a cheaper, simpler alternative. Once users are locked into an ecosystem, the switching costs create a powerful barrier, tying them to a platform and pulling in even more users in a self-reinforcing cycle.

Quantum leaps in business model evolution

Technology is driving significant changes in business models, quickly dismantling existing approaches. Companies are forced to transform at a pace that traditional change management cannot handle. Digital transformation, in particular, is not an incremental process. For Quantaland Media, the shift from print to digital was a sudden and drastic change, requiring a swift and unsettling transition period.

The platform economy triggered a quantum displacement of intermediaries. Digital platforms enabled direct connections, sidelining travel agents, stockbrokers, and real estate agents almost overnight. Similarly, ride-sharing models created quantum gains in asset utilization. Quantaland Transport's fleet efficiency soared exponentially after they started using a digital platform that matched demand dynamically. This level of growth was impossible to achieve under traditional dispatch systems.

On-demand services shifted customer expectations instantly, demanding instant fulfillment and forcing rapid operational overhauls. Now, AI is sparking the next great leap. It incrementally automates complex tasks in customer service, financial analysis, and logistics once its capabilities cross the threshold of human competency.

Recognizing this quantum nature is essential. The speed and scale of technological change will always blindside linear long-term planning. The only successful strategy is to build flexibility and optionality into every decision, preparing for innovations that can transform entire industries in the time it takes to analyze a quarterly report.

Meanwhile, far from Quantaland

Silicon Valley Bank collapse (2023)

The Quantum State: Silicon Valley Bank's (SVB) investment portfolio existed simultaneously as both safe and risky. This superposition collapsed into failure when rising interest rates exposed the mismatch between accounting treatment and economic reality.

The Story: The collapse of SVB in March 2023 is a textbook example of how the illusion of long-term stability can quickly transform into financial catastrophe. SVB specialized in serving technology startups, venture capital firms, and high-net-worth individuals in the tech sector.

SVB's downfall stemmed from a seemingly conservative investment strategy that proved disastrously flawed. During 2020–2021, the bank experienced massive deposit inflows as deposits from technology companies grew in an unprecedented manner in the face of pandemic-related markets. SVB's deposits more than tripled from $62 billion at the end of 2019 to over $190 billion by the end of 2021. Facing limited loan demand and pressure to generate returns, SVB invested much of this excess cash in long-duration securities, favoring U.S. Treasuries and MBS. By the end of 2022, the bank had approximately $91 billion in held-to-maturity securities with an average yield of just 1.56%. These investments were safe from a credit standpoint but carried hidden interest-rate risk. SVB classified the majority of these securities as "held-to-maturity" rather than "available-for-sale." This allowed them to avoid marking these investments to market at a time when treasury rates kept increasing. As of December 2022, SVB's securities portfolio had approximately $15 billion in unrealized losses that weren't reflected on its balance sheet. Just to put the figure in perspective, this was nearly equal to the bank's total equity capital.

The situation deteriorated rapidly in early 2023. Venture capital funding dried up, and tech companies began drawing down deposits to fund operations. SVB had to sell securities to meet withdrawal demands. On March 8, 2023, the bank announced it had sold $21 billion of securities at a $1.8 billion loss and was planning to raise $2.25 billion in capital to support its balance sheet.

This announcement triggered panic among depositors and investors. SVB sought to reassure their customers and advised them not to panic. This ill-conceived statement had precisely the opposite effect. Within 24 hours, customers attempted to withdraw $42 billion. A bank run is not a new phenomenon, but the speed it attained with digital banking and social media was unexpected. By March 10, 2023, regulators closed SVB and placed it into Federal Deposit Insurance Corporation (FDIC) receivership. This was the second-largest bank failure in U.S. history and the largest since the 2008 financial crisis.

The contagion quickly spread to other regional banks with similar characteristics, such as First Republic, PacWest, and Western Alliance, which experienced severe deposit outflows and stock price declines. To contain the crisis, U.S. regulators announced extraordinary measures. On March 12, they guaranteed all deposits at SVB, even those above the $250,000 FDIC insurance limit. They also established a new Bank Term Funding Program that allowed banks to borrow against their underwater bond portfolios at par value. Poor management of interest-rate risk and overt reliance on uninsured deposits contributed to the fall.

The Quantum Finance Lesson: The collapse of SVB is a clear example of a quantum paradox in risk assessment. Depending on the analytical method applied, assets appeared simultaneously secure and perilous. Much like a quantum particle that holds several possible states until measured, SVB's investment holdings existed in a superposition of safety and danger. The credit risks were minimal, but interest-rate exposure posed a significant threat.

Accounting conventions concealed this complexity, hiding the true risks, until a crisis forced a sudden confrontation with the truth. In quantum mechanics, it is well accepted that the opposite of one truth is not necessarily falsehood. It can represent an equally valid truth that emerges from a change in perspective and is valid within the new context.

Practical Application: Financial risk managers should recognize that different analytical frameworks reveal complementary aspects of risk rather than definitive assessments. Incorporating quantum thinking would help avoid overreliance on any single risk measure and accept uncertainty as an inseparable part of reality. This quantum perspective acknowledges that financial stability depends on probability distributions rather than deterministic factors.

Kodak's decline

The Quantum State: Kodak existed in a superposition of market leader and obsolete business, with digital technology gradually collapsing this duality toward obsolescence. This happened despite the company's technical capability to adapt.

The Story: Eastman Kodak's journey from a global photography giant to bankruptcy is a stark lesson in the illusion of long-term stability. A technological paradigm shift shattered its apparent invincibility. Founded in 1888 by George Eastman, Kodak dominated the photographic film market for most of the 20th century. It achieved a 90% market share in the United States by the 1970s and employed over 145,000 people at its peak.

Ironically, Kodak itself invented the core technology that would eventually destroy its business model. In 1975, Kodak engineer Steven Sasson created the first digital camera prototype. It was a toaster-sized device that captured black-and-white images at 0.01 megapixels. Kodak, obsessed with photographic film, saw this as a threat to their own business model. Kodak saw the potential of digital photography but wasn't sure when the market would shift to it. This uncertainty led them to hedge. Instead of fully embracing digital transformation, they split investments between digital capabilities and their highly profitable Q$ film business. Through the 1990s, Kodak made several digital forays, including launching the

PhotoCD format and developing early consumer digital cameras. But all these initiatives were consistently subordinated to protecting the film business, which still generated over 70% of the company's revenue. Kodak poured billions into film manufacturing capacity well into the 2000s, even as digital camera sales began to accelerate. By 2003, digital camera sales surpassed film camera sales in the United States, and by 2011, Kodak's film business had collapsed to less than 10% of its peak volume.

Kodak tried new business models to offset the collapsing film business. Their investments included consumer printers, pharmaceutical ingredients, and digital image licensing. None gained enough traction to replace the declining film revenue. In January 2012, Kodak filed for bankruptcy protection under Chapter 11 of the United States Bankruptcy Code with $6.75 billion in assets and $7.3 billion in liabilities. When Kodak emerged from bankruptcy in September 2013, it had drastically changed, employing just 8,000 people and concentrating on commercial printing technologies.

Kodak's decline was striking despite its technological expertise and financial resources to pivot to digital imaging. They even saw digital photography's potential but underestimated its overwhelming speed and scale.

The Quantum Finance Lesson: Kodak's decline highlights the quantum nature of business evolution. Companies, much like quantum systems, can exist in a superposition of stability and obsolescence. A technological revolution forces those possibilities to collapse into a single outcome. Just as measuring a quantum state causes a wave function to collapse, the rise of digital technology ultimately shattered Kodak's longstanding dominance.

The concept of quantum decoherence, where systems lose their wave-like ambiguity through interactions with their environment, applies here as well. As digital technology reshaped the landscape of photography, Kodak's film-based business model lost its coherence and relevance. Unable to shift fully to the new reality, Kodak's legacy business gave in as consumers and professionals embraced digital solutions. When the broader environment moves toward a new paradigm, resistance eventually leads to irrelevance or exit.

Practical Application: Business leaders should recognize that long-term stability is always probabilistic rather than deterministic. This means you should continuously experiment with potential future states rather than optimize exclusively for current conditions. This quantum perspective acknowledges that business success requires maintaining superpositions of current and future capabilities. Excelling in one area can blind you to disruptive alternatives, even those you develop internally.

Blockbuster versus Netflix

The Quantum State: The home entertainment market existed in a transitional state between physical and digital distribution. The reigning champions, Blockbuster, failed to maintain flexibility across these competing paradigms, while Netflix succeeded in navigating both simultaneously.

The Story: The contrasting fates of Blockbuster and Netflix show how the illusion of business stability can blind established companies to emerging paradigm shifts. Founded in 1985, Blockbuster grew to become the dominant player in video rental, operating over 9,000 stores worldwide and generating $5.9 billion in revenue at its peak in 2004. Blockbuster's business model depended on physical DVD rentals, late fees (which constituted between 10% and 20% of its revenue), and prime retail locations. This model had proven remarkably stable and profitable for nearly two decades, creating a sense of permanence that proved dangerously misleading.

Netflix, founded in 1997 by Reed Hastings and Marc Randolph, initially competed with Blockbuster through DVD-by-mail rentals. Blockbuster could have easily adopted the model but ignored it. In fact, in 2000, Netflix proposed a partnership to Blockbuster where Netflix would manage Blockbuster's online brand while Blockbuster would promote Netflix in its stores. The proposal was shot down by Blockbuster, reflecting their confidence in the stability of their business model. In 2007, the landscape was clearly shifting. Netflix launched its streaming service, while Blockbuster belatedly introduced its own DVD-by-mail program. By 2009, Netflix had 10 million subscribers, while Blockbuster was struggling with declining store traffic and mounting debt from a leveraged buyout.

The contrast in strategic flexibility became increasingly apparent. Netflix maintained its DVD-by-mail service and simultaneously invested in streaming. On the other hand, Blockbuster oscillated between trying to compete online and doubling down on its store-based model. This failure to recognize the quantum nature of the evolving video market proved fatal. Netflix understood it was operating in a transitional period and maintained DVD services while building streaming capabilities. Blockbuster held onto the belief that physical rentals would either maintain their dominance or face rapid replacement.

By 2010, Blockbuster had filed for bankruptcy protection under Chapter 11 of the United States Bankruptcy Code with nearly $1 billion in debt. At the time, Netflix had 16 million subscribers and was rapidly expanding its streaming catalog. The final Blockbuster store in the United States operated as a franchise. This serves as a poignant and sentimental remembrance of a once-dominant business model.

Netflix, meanwhile, evolved from content distributor to creator, launching its first original series, House of Cards, in 2013. By 2023, Netflix had over 230 million subscribers worldwide and had fundamentally transformed the entertainment industry. The contrasting trajectories reflect different approaches to uncertainty. This quantum approach to business development allowed Netflix to adapt as the probability landscape evolved.

In contrast, Blockbuster remained trapped in a deterministic mindset that proved incompatible with technological reality.

The Quantum Finance Lesson: The experiences of Blockbuster and Netflix vividly exemplify the quantum nature of market evolution. Industries often balance between competing technological paradigms, making it essential for companies to remain adaptable and open to multiple future paths. Much like quantum particles that can follow several trajectories at once, businesses must be ready to explore various strategic directions during times of transformation.

The quantum concept of tunneling explains Netflix's approach, as it allows particles to pass through barriers that classical physics considers impenetrable. By maintaining both physical and digital distribution models throughout the transition, Netflix managed to navigate beyond the perceived limitations of each. By carefully observing early signs pointing to digital distribution's potential, Netflix was able to adapt progressively to the changing market landscape.

Practical Application: Finance leaders should embrace quantum thinking, keeping strategic options open during technological shifts and avoiding premature all-or-nothing choices. This means investing in both current and emerging models until the market reveals the best path. This quantum approach acknowledges that business evolution, like quantum systems, follows probability waves rather than deterministic paths. This needs flexibility and patience to navigate successfully.

Looking ahead

The illusion of long-term stability has cost countless businesses and investors dearly, from Kodak's digital blindness to pension funds caught by duration mismatches. Yet recognizing uncertainty as fundamental rather than problematic opens a powerful new path forward. Instead of betting everything on a single predicted future, quantum finance teaches us to prepare for multiple scenarios simultaneously. This preparation isn't about hedging against failure. Instead, it's about positioning for success across various possible outcomes. In our next section, we'll discover how the most successful financial decision-makers don't choose between competing strategies. They actually hold multiple approaches simultaneously, maintaining what quantum physics calls superposition until market conditions reveal the optimal path.

Chapter 8

Financial superposition: holding multiple perspectives

"*The crane knows when to stand on one leg and when on both*," goes an old Quantaland proverb.

Portfolio manager Sarah stared at her screen in disbelief. Her carefully constructed "growth-focused" portfolio was hemorrhaging value as interest rates climbed in early 2022. Down the hall, her colleague Marcus was celebrating as his "value-oriented" strategy was finally paying off after years of underperformance. Both were smart, experienced professionals using sound investment principles. Yet in the same market, one was winning while the other was losing. Sarah's mistake wasn't her analysis. It was her assumption that she had to choose sides. This chapter reveals why the most successful financial decision-makers don't pick between competing strategies and hold multiple approaches simultaneously like quantum particles existing in several states at once. You'll discover how Bridgewater's All Weather Portfolio thrives by embracing contradictions, learn from Slack's masterful pivot that maintained dual identities, and understand why the best investors treat uncertainty not as a problem to solve but as a resource to exploit. By the end, you'll see how holding opposing positions simultaneously can transform market volatility from your enemy into your greatest strategic advantage.

The strategic superposition: managing multiple futures at once

Financial superposition is the practice of consciously maintaining multiple, valid strategic approaches to a single decision at the same time. Instead of forcing a premature choice based on inevitably incomplete information, this method allows market conditions to evolve and reveal which strategy will be most effective. It is a formal recognition that financial markets – much like quantum systems – frequently exist across a spectrum of potential states. They only collapse into a single reality when a specific

DOI: 10.4324/9781003619635-8

event, such as an earnings report, a central bank decision, or a geopolitical shock, acts as a measurement, forcing a definitive outcome.

This represents a significant shift from traditional financial doctrine. Conventional wisdom prizes decisive action, celebrating the leader who analyzes the data, forms a conviction, and commits a move. Quantum finance does not dismiss this outright but reveals that it comes with a hidden cost. Premature decisiveness can systematically eliminate valuable optionality and deprive an organization of potentially more effective paths that emerge later. By deliberately holding strategies in superposition, financial decision-makers preserve their flexibility. They build a portfolio of strategic options, positioning themselves to adapt with agility as the future reveals itself. This is a better option than betting everything on a single forecast of what the future will be.

Running in multiple lanes: the power of parallel strategies

Financial superposition empowers organizations to pursue multiple investment strategies in parallel, rather than betting everything on a single approach and hoping it proves correct. This provides a flexible approach, enabling organizations to choose the best strategy for the current market situation without having to completely rebuild their investment portfolio, which can be costly and time-consuming.

Maintaining multiple investment approaches simultaneously

Modern investment management uses quantum superposition by letting portfolios include different strategies at the same time, similar to how particles can be in several states. Quantaland Asset Management operationalizes this by simultaneously balancing growth-focused tech stock investments with value-oriented traditional stocks. Instead of choosing one over the other, they let market conditions reveal which strategy will dominate. When technology stocks surge in growth potential, their value holdings might decline due to investor rotation. However, the overall portfolio benefits from the net gain rather than completely missing the opportunity.

This moves beyond traditional diversification. While traditional diversification spreads risk across different asset classes, quantum-inspired strategies deliberately pair opposing positions within the same markets. This strategy is designed to profit regardless of directional moves in underlying assets. It is the difference between checking the weather forecast and simply carrying both an umbrella and sunglasses ready for any outcome.

Factor investing takes this further, capturing both momentum and mean-reversion premiums simultaneously, even when their signals conflict.

The strategy keeps both factors active until market conditions show which one will lead. Geographic superposition operates on a similar principle, balancing exposure to emerging markets for growth and developed markets for stability, without attempting to time the switch perfectly between regions.

The key insight is that financial superposition doesn't try to eliminate contradiction; it embraces it as a fundamental feature of uncertain markets. The portfolio, like a quantum system, holds multiple states until a market event acts as a measurement, collapsing the possibilities and revealing a winner.

Portfolio diversification beyond asset classes

Quantum thinking pushes diversification beyond spreading capital across asset classes. It expands to include multiple strategies, time horizons, and behavioral approaches all operating at once, letting market conditions determine which strategy is most effective. This principle may manifest in various ways.

Let's look at the time-horizon superposition strategy. Under this strategy, portfolios hold short-term tactical positions to capture quick opportunities, along with long-term strategic positions for secular trends. This allows them to profit from transient market inefficiencies while maintaining exposure to enduring values without requiring perfect timing.

Similarly, one can have behavioral diversification-inspired strategies. Here, one can simultaneously run momentum-based approaches that chase trends alongside contrarian ones that bet on reversals. This recognizes that markets exhibit multiple behavioral patterns at the same time.

Quantaland Hedge Fund puts them into action, running long-short equity, merger arbitrage, and volatility trading strategies concurrently. Each strategy responds differently to a given shock like a sharp rate spike might hurt one while benefiting another.

This extends to sophisticated volatility and currency strategies. A portfolio can hold both long and short volatility positions, gaining whether market volatility rises or falls without needing to predict its direction. Similarly, currency superposition combines direct forex speculation with international investments that protect against currency fluctuations. This allows investors to capture potential currency moves separately from the performance of the foreign assets themselves.

The core insight is that true diversification succeeds when you embrace and maintain multiple valid approaches simultaneously. You don't need to try to pick the one "right" strategy in advance.

Strategic hedging and contradiction management

Superposition strategies deliberately employ contradictory positions that may seem illogical but provide protection against multiple outcomes at once. This requires accepting apparent inconsistencies that traditional finance often avoids. You can see this in practice in various strategies.

The interest-rate contradiction management approach uses duration laddering to maintain exposure to both rising and falling rate environments. Inflation hedging involves blending real assets (like commodities), inflation-protected securities, and nominal bonds to maintain resilience across any inflation scenario. Similarly, investors manage credit and market direction contradictions by holding corporate bonds for yield and simultaneously purchasing CDS for protection. Similarly, long-short equity strategies aim to profit from security selection regardless of whether the broader market rises, falls, or moves sideways.

Quantaland Insurance employs a sophisticated version of this by acting as both a buyer of traditional catastrophe reinsurance and an investor in catastrophe bonds (cat bonds). While this appears to place them on both sides of the same risk, it is a deliberate portfolio optimization technique. The traditional reinsurance protects their core business from catastrophic losses. Conversely, their cat bond investments represent a source of potential underwriting profit from unrelated risks, diversifying their income streams. This structure mitigates their need to perfectly predict the frequency of major events. A key advantage is the mitigation of counterparty risk. Under this strategy, their traditional coverage is exposed to the solvency of other firms, while their positions in fully collateralized cat bonds are secured by pre-funded capital.

These seemingly contradictory positions build resilience by preparing for multiple realities at once, rather than betting on a single outcome.

Dynamic strategy allocation based on market conditions

The most advanced application of superposition uses dynamic allocation systems that automatically adjust the weight given to various strategies based on real-time market conditions. This simulates a continuous quantum measurement process.

These systems do not predict the best strategy. Instead, they hold multiple approaches at once and adjust emphasis as market conditions evolve and patterns emerge. This mirrors quantum systems remaining in superposition until measurement reveals the dominant state.

For example, volatility-based allocation shifts emphasize momentum strategies during high-volatility trends and move to mean reversion during calm, range-bound periods. Correlation-based allocation monitors how asset classes move together. This deliberately reduces reliance on diversification benefits during crisis periods when correlations spike and hedging activities increase.

Economic regime-driven allocation tracks macroeconomic indicators (e.g., GDP growth, inflation) to emphasize cyclical strategies during expansions and defensive ones during recessions. However, exposure is maintained throughout both scenarios.

Market microstructure-driven allocation monitors trading patterns, liquidity, and order flow to determine whether high-frequency or fundamental strategies are more suited to current conditions. A sentiment-based allocation system gauges investor psychology to shift between contrarian approaches during periods of extreme sentiment and momentum strategies during periods of moderate, trending conditions.

Inspired by this quantum view, Quantaland Trading uses algorithmic systems that continuously measure these market conditions. The algorithms then quickly choose how to switch between trend-following, mean-reversion, and volatility-arbitrage strategies, making sure the portfolio matches the current market situation.

The option mindset: buying the right to a better future

Optionality is the practice of building strategies that maintain multiple potential futures simultaneously. Instead of making premature, irreversible choices, organizations create the right, but not the obligation, to act. Now one can wait for specific events to reveal the most advantageous path. This approach converts uncertainty from a problem to be eliminated into a resource for management and exploitation.

Real options in corporate finance

Real options theory applies the principles of financial options to corporate strategy. This approach recognizes that certain investments create valuable future opportunities beyond their immediate, measurable returns.

Quantaland Manufacturing embodies this by building modular production facilities. This creates a clear real option by incurring a minor extra cost for flexibility today. This grants them the right, but not the obligation, to expand rapidly if demand grows. The payoff is asymmetric, with the limited downside of the initial cost and unlimited upside of capturing new demand.

This mindset applies to numerous corporate functions. In R&D strategy, each project is a call option on a future market. Most may expire worthless, but a single success can justify the entire portfolio. In the case of an acquisition strategy, this will involve forming partnerships and taking minority stakes in startups. Quantaland Technology follows this strategy, and their investment is akin to effectively paying a premium for a call option to acquire the company later if its technology proves viable.

In cases of international expansion, using pilot programs and joint ventures creates a low-cost option to enter a market, which will be exercised only if conditions are promising. Investment in cloud infrastructure by technology platforms or having flexible manufacturing platforms creates an option for multiple future product lines or services without being locked into one.

Real options exist in a state of superposition. They are simultaneously valuable and worthless until market conditions collapse them into a specific outcome. They derive their value from their asymmetric payoff profile, where success yields significant returns and caps losses at the initial investment.

Financial options as a uncertainty management tool

Financial options are one of the purest instruments for harnessing uncertainty. They provide asymmetric payoffs, offering significant benefits from favorable price movements while strictly limiting losses to the premium paid.

This is the essence of portfolio insurance. Strategies like the collar, involving selling call options and buying puts around an equity position, establish performance boundaries. They protect the downside while preserving most of the upside, acknowledging that market outcomes exist in superposition until a price movement locks in a result.

This philosophy applies across markets. In currency hedging, options protect against adverse FX movements while preserving the benefit from favorable shifts. In contrast, forward contracts lock in a rate and eliminate both risk and opportunity. Bond portfolios are protected from rising yields by options for managing interest rates, which also allow investors to keep profits in the event that yields decline. In volatility trading, strategies such as straddles and strangles directly profit from large price movements in either direction. They are a pure play on uncertainty itself, removing the need to predict market direction.

The power of options is that they don't try to eliminate uncertainty; they weaponize it, creating value from the very unpredictability that traditional risk-management fears.

Strategic flexibility and reversible decisions

Quantum thinking promotes strategic flexibility by favoring reversible decisions and avoiding irreversible commitments until the last responsible moment. This is the art of maximizing optionality by keeping future alternatives alive.

Using contractors and temporary workers achieves operational flexibility. This allows flexibility to scale the workforce up or down without any hiring commitment. Partnership structures like joint ventures and alliances are strategic options; they open doors and create opportunities without the finality of a merger or acquisition.

Quantaland Retail prefers flexible leases with expansion and contraction clauses over purchasing property. This gives them the right to adapt their footprint based on performance and avoids the long-term commitments associated with acquiring physical assets.

Technology and financial flexibility come with the use of modular, cloud-based systems, which are options for future IT needs. Similarly, maintaining multiple lines of credit is an option for future funding opportunities. Multi-sourcing is an option that brings flexibility to the supply line. It avoids the irreversible commitment and vulnerability of relying on a single supplier.

This flexibility adds value by maintaining a superposition of strategic paths. Like quantum particles, these approaches hold multiple valid states until market conditions force a measurement and a choice, dramatically boosting an organization's adaptive capacity.

Option valuation beyond Black-Scholes

Valuing these real options requires moving beyond the simplified assumptions of the Black-Scholes model to capture the complex, messy reality of business. Traditional analytical models fall short because business options are path-dependent and interact with competitive and managerial behavior. Advanced methods are needed:

Consider the tool of decision tree analysis. This tool maps out the sequence of potential decisions and their consequences, incorporating the value of choices at each juncture. Monte Carlo simulation generates thousands of potential scenarios to model the range of possible outcomes and their probabilities, capturing complexities that analytical models overlook by design.

Behavioral valuation is another interesting application. This approach recognizes that biases, risk appetite, and organizational constraints influence management decisions, which are not purely economically rational. Using Game theory in valuation opens up another dimension. This tool could be used for valuing similar options held by competitors. This is useful when the value of your option to enter a market depends critically on the timing and strategic moves of your rivals. Fuzzy logic-driven valuation deserves careful consideration. This approach embraces ambiguity when precise probabilities cannot be determined. It uses approximate reasoning to handle the inherent vagueness in many business parameters.

The key insight is that business options exist in complex superposition states. Valuation models that force them into oversimplified, deterministic frameworks will inevitably fail to capture their true strategic worth. Proper appraisal requires methods that embrace, rather than ignore, uncertainty and complexity.

Rehearsing reality: using scenarios to stress-test your strategy

Scenario analysis and contingency planning are systematic engines for managing quantum uncertainty. Instead of betting on a single predicted outcome, these tools allow organizations to explore and prepare for multiple potential futures simultaneously. They formally acknowledge that the future exists in a superposition of states, only collapsing into a single reality when specific events act as measurements. These practices provide the framework for making decisions that remain robust across a spectrum of possibilities.

Stress testing

Stress testing is the deliberate application of quantum measurement to financial systems. By applying extreme but plausible scenarios, these tests reveal hidden vulnerabilities and unexpected correlations that remain entirely invisible under normal, stable conditions. They function like forced observation, collapsing the complex, entangled relationships within a business or portfolio into specific, measurable, stress-induced responses.

Regulatory stress-testing requirements

For banks, mandatory stress tests are definitive quantum measurement events imposed by regulators. The prescribed scenarios don't just assess existing strength; they actively reveal hidden weaknesses and force capital adjustments, effectively shaping the future behavior and risk profile of the institution.

These requirements force banks to maintain capital buffers that may seem excessive in advantageous times but prove essential under stress. The publication of results creates a powerful observer effect, directly influencing market perceptions and funding costs. This often leads banks to optimize their strategies for expected test results rather than achieve pure economic efficiency.

Quantaland Bank's annual comprehensive capital analysis and review serves as this exact type of observation. The "severely adverse" scenario generates conditions designed to break normal correlations and uncover concentrations that benign markets hide. The test, apart from measuring risk, forges a new risk-management reality by forcing preparation for scenarios that transform normal assumptions into crisis states.

Internal stress testing for strategic planning

Forward-thinking organizations build upon regulatory requirements by designing their own, often more severe, internal stress tests for strategic planning.

For instance, Quantaland Manufacturing conducts stress tests on its entire business model to withstand the potential impact of supply-chain disruption, demand collapse, and spikes in input costs. These probe for dependencies and identify single points of failure that standard planning completely misses.

This internal measurement extends across various business functions. Revenue stress testing examines what happens when core business assumptions fail simultaneously. Liquidity testing focuses on cash sustainability when funding markets freeze, and working capital needs surge. Operational testing probes resilience against technology failure, critical personnel loss, or external shocks like cyberattacks. Market and ESG stress tests assess sustainability amid dramatic competitive shifts or challenges from regulation and stakeholder activism.

Like a quantum measurement, internal stress testing collapses strategic assumptions into specific scenarios, revealing which choices remain viable when the environment turns adverse.

Reverse stress testing and snapping point analysis

Reverse stress testing is a more radical form of quantum measurement. Instead of applying stress to see what happens, it works backward from a predetermined outcome like business failure to identify the scenarios that could cause it, regardless of their perceived probability. This approach, which includes breaking point analysis, pinpoints absolute thresholds for failure that probability-based models often ignore.

Reverse stress testing can be used in various areas of business. The revenue breaking point measures the minimum performance level required for financial sustainability. Capital breaking point defines the minimum capital level needed to survive extreme stress. Similarly, the operational breaking point computes the minimum capacity required for business continuity, while the liquidity breaking point identifies the cash scenarios that would force emergency asset sales or funding.

Quantaland Insurance uses this to ignore historical frequency and identify catastrophic scenarios that could threaten solvency. This approach would uncover concentration risks and reinsurance needs that conventional models underestimate. This method focuses on absolute vulnerability, free from the constraints of potentially flawed probability distributions.

Dynamic stress testing with real-time adjustments

An advanced application is dynamic stress testing, which creates a quantum-like system that evolves in real time with the business environment. This approach addresses the limitations associated with relying on outdated, static, and periodic scenarios.

Quantaland Technology does this by constantly watching and changing its stress scenarios based on real-time signs of market changes, competition, and technology shifts. This provides a real-time view of resilience.

ML enhances this further by identifying hidden patterns and correlations, generating unique stress scenarios that are often beyond human design. An integrated approach blends regulatory requirements, strategic internal tests, and dynamic monitoring into a comprehensive system.

This forward-looking method treats risk not as a set of fixed scenarios but as an evolving probability field. It allows for proactive preparation, embodying the quantum principle that the act of observation must adapt as the system itself changes.

Mapping the probability landscape: the Monte Carlo advantage

Monte Carlo simulations provide the ultimate computational embodiment of quantum-inspired thinking in finance. By generating thousands of random scenarios, they explore the entire spectrum of potential outcomes rather than relying on a flawed, single-point estimate. These methods don't try to eliminate uncertainty; they embrace it as a fundamental characteristic of the system, allowing decision-makers to navigate it intelligently.

Understanding probabilistic outcomes

The power of Monte Carlo simulation lies in its ability to reveal full probability distributions without complex mathematics. It reflects quantum-like thinking by generating and analyzing thousands, or even millions, of potential futures simultaneously. By varying all uncertain parameters at once, it captures crucial interaction effects and correlations that traditional, one-at-a-time sensitivity analysis completely misses.

This approach makes uncertainty tangible for nontechnical decision-makers through intuitive risk visualizations and scenario rankings that identify the combinations of factors with the greatest impact. Confidence-interval analysis translates probabilistic outcomes into business terms, offering a quantum-like view of what is certain and what remains uncertain.

Quantaland Pension Fund uses this method to assess its liability matching across a full range of economic scenarios, moving far beyond the false security of assuming fixed discount rates. The simulation portrays business outcomes as existing in a superposition of potential states, only crystallizing into a single result once the future unfolds.

Monte Carlo applications in portfolio management

In portfolio management, Monte Carlo simulation models the complex interaction among multiple uncertain factors to optimize real-world conditions. It has multiple applications.

In the case of asset allocation, it finds portfolio compositions that maximize the probability of achieving client objectives across varied market environments. This effectively sidesteps the false precision of traditional mean-variance optimization. For retirement planning, it builds strategies that factor in longevity risk, uncertain returns, and inflation acting together, not in isolation.

Applied to risk budgeting and attribution, it allocates risk based on contributions to overall portfolio traits. This approach distinguishes skillful alpha from random luck by analyzing the likelihood of results occurring by chance.

Quantaland Asset Management runs thousands of market scenarios to identify strategies with the highest probability of success for their clients' goals. This approach fundamentally views investment returns as probability distributions and not fixed expectations. This encourages a mindset that explores multiple market realities until conditions eventually pick a winner.

Limitations and assumptions in simulation models

However, Monte Carlo simulations possess their own quantum-like limitations. The model's output is heavily influenced by its input assumptions and design, creating a form of observer effect where the measurement tool shapes the measured reality.

"Garbage in, garbage out" is supplemented by "garbage assumptions, garbage strategies" as a fundamental law. Poor assumptions about return distributions, volatility, or, more critically, correlations will produce misleadingly precise output. Often, these models make the fatal assumption of correlation stability, which can lead to a spike toward 1 during a crisis. A critical error in a key assumption led to the failure of Quantaland Financial Services' credit models.

Path dependency is another critical factor. Outcomes often depend on the sequence of events, not just on their final values. Despite their sophistication, these models do not reveal the objective truth. They produce realities shaped by the modeler's choices much like a quantum experiment. Careful validation against historical stress periods is essential to keep these assumptions reasonable and to avoid a false sense of security.

Behavioral Monte Carlo incorporating human decision-making

The most advanced application of this technology incorporates the quintessentially human element – the behavioral biases and decision-making under pressure. This recognizes that outcomes are not purely statistical, and they are primarily determined by how people react.

Advanced simulations can now model various aspects of behaviorally influenced decision-making. The management responses model can examine how leadership teams make decisions under stress, which often deviate radically from calm-era planning assumptions. Studying competitive and regulatory reactions reveals how rivals and regulators will change their policies and strategies in response to adverse conditions.

Customer and investor psychology-focused simulations show how client and investor behavior, such as panic, euphoria, herding, and so on, impacts funding costs, redemptions, and demand. The range of outcomes is often beyond what fundamentals would suggest.

Quantaland Corporation uses behavioral simulation to model management decision patterns under pressure, customer reactions to service disruptions, and investor psychology. This captures the quantum-like interplay between external conditions and internal responses that pure statistical models entirely overlook.

Meanwhile, far from Quantaland

Technology firms during the 2022 downturn

The Quantum State: Leading technology companies maintained superpositions of growth and efficiency. They simultaneously pursued cost-cutting measures and strategic investments as the market transitioned from expansion to consolidation.

The Story: The technology sector's response to the 2022 market downturn offers a clear example of financial superposition in action. A sharp reversal in 2022 ended the pandemic-fueled boom that drove tech valuations to record highs. Rising interest rates, inflation concerns, and growth slowdowns triggered the worst year for the NASDAQ since 2008. The index fell 33.1%.

Facing this tough environment, leading tech firms showed strategic flexibility by pursuing contradictory initiatives. They cut costs aggressively while continuing to invest in strategic growth areas. This dual approach allowed them to address short-term market concerns while maintaining long-term competitive positioning.

Let us see how Google parent Alphabet used this approach. In January 2023, Alphabet announced layoffs of about 12,000 employees, roughly 6% of its global workforce. These cuts focused primarily on nontechnical roles and projects deemed nonessential, generating immediate cost savings and improved efficiency metrics. At the same time, Alphabet increased R&D investment in AI and cloud computing. These are the areas Google deemed strategically vital. This dual focus on efficiency and investment represented a classic superposition strategy, allowing Alphabet to navigate competing market pressures.

Microsoft demonstrated similar strategic duality. In January 2023, the company announced 10,000 layoffs while simultaneously revealing plans to invest $10 billion in OpenAI, the company behind ChatGPT. This approach balanced appeasing Wall Street's immediate concerns about profitability with positioning for future growth opportunities. Let's look at Amazon. Their response further illustrates this quantum-like approach. The company laid off over 18,000 employees while continuing major investments in the logistics infrastructure and AI capabilities.

You would find it interesting to review how these companies communicated their dual strategies to different stakeholders. To investors, they emphasized cost discipline and margin improvement. Microsoft highlighted that its layoffs would result in $1.2 billion in cost savings. Yet to customers and developers, they stressed continued innovation and expanded capabilities. Amazon emphasized that Amazon Web Services (AWS) continued to grow its engineering teams despite corporate layoffs.

This strategic superposition proved effective. By Q3 2023, all three companies had reported improved profit margins while maintaining competitive positions in emerging technologies. This balancing act was often beyond the financial and strategic capabilities of smaller firms.

The Quantum Finance Lesson: The tech sector's approach to the downturn serves as a clear example of financial superposition. Companies managed to keep several strategic options open at the same time, postponing a definitive course of action until the market clarified which strategy would succeed. Just as quantum particles exist in several possible states until observed, organizations facing uncertainty can pursue apparently conflicting paths simultaneously.

This mirrors the quantum theory of wave-particle duality, where entities may present as either waves or particles depending on how they're measured. In the same way, during this period, technology firms positioned themselves as both growth and value investments. Their public messaging was tailored according to the audience and context, highlighting different aspects as needed.

Similar to superposition discussions, this situation demonstrates how the questions posed influence the outcomes. Technology companies recognized that different stakeholders "measure" their performance from distinct viewpoints, and so they maintained a flexible strategy to address these various perspectives at once.

Practical Application: Business leaders should incorporate quantum thinking by maintaining strategic optionality during periods of uncertainty. This means simultaneously pursuing efficiency improvements and strategic investments. This approach proves superior to making binary choices that prematurely commit to a single market narrative.

Tech firms recognized their constraints and opportunities and acted to address both. This quantum perspective highlights that business strategy, like quantum systems, often involves holding multiple potential states until market evolution brings clarity.

Investment portfolios with diverse assets

The Quantum State: Diversified investment portfolios maintain positions across multiple asset classes that respond differently to various economic scenarios. This approach creates a superposition of strategic postures that enhances resilience against uncertainty.

The Story: Ray Dalio's All Weather Portfolio represents one of the most influential applications of quantum-like thinking to investment strategy. Dalio, the founder of Bridgewater Associates, the world's largest hedge fund, developed this approach in the 1990s. The All Weather approach recognized that financial markets are unpredictable and created investment portfolios that can do well in different economic situations.

The All Weather strategy stemmed from Dalio's recognition that traditional portfolio theory fell short on true market uncertainty. It focused on assets that perform well in different environments, ranked by their ability to handle various conditions. This approach closely resembles quantum thinking by maintaining multiple strategic positions simultaneously rather than betting on a single economic forecast.

The traditional All Weather Portfolio allocates assets across four economic scenarios: rising growth (30% equities), falling growth (40% long-term bonds), rising inflation (15% intermediate-term bonds), and falling inflation (15% commodities, 7.5% gold, and 7.5% other commodities). This structure holds a superposition of four economic possibilities, with each component designed to perform well in a specific environment. The overall portfolio maintains resilience regardless of which scenario materializes. The 2008 financial crisis dramatically tested the All Weather approach, resulting in major losses for most investment strategies. In contrast, Bridgewater's more active strategies demonstrated exceptional performance. Their Pure Alpha fund booked a return of 9.5% in 2008 when the S&P 500 lost 37%. The passive All Weather approach also demonstrated remarkable resilience. Although the passive All Weather approach lost about 4%, it still outperformed traditional balanced portfolios.

This outperformance continued through subsequent crises. During the COVID-19-induced market shock in Q1 2020, when the S&P 500 fell 20%, an All Weather-style portfolio declined just 1.9%. This resilience comes from embracing uncertainty in a quantum-like manner rather than predicting specific outcomes. Other investors have adopted similar quantum approaches. David Swensen, who managed Yale's endowment from 1985 until his death in 2021, pioneered the "Yale Model" of institutional investing. This model spreads assets across conventional investments, private equity, real assets, and absolute return strategies. Swensen acknowledged the impossibility of consistent market timing. His focus was on building portfolios that perform acceptably in all environments.

These quantum approaches differ significantly from traditional diversification. Conventional portfolios might diversify across stocks and bonds based on risk-return relations. True quantum portfolios select assets for their differing responses to various economic conditions. The strategy doesn't focus on beating the market. It accepts market uncertainty and prepares for it.

The Quantum Finance Lesson: The All Weather investment approach exemplifies financial superposition in portfolio design. Rather than making an all-or-nothing commitment to a single forecast, this strategy allows investors to maintain exposure to multiple economic scenarios at once. Diversified portfolios, like quantum particles, can hold various economic positions until real events reveal which one will be the most accurate.

This strategy mirrors the quantum principle of complementarity, where different methods of measurement reveal unique facets of reality. In constructing the All Weather Portfolio, assets are chosen for their varied responses to factors such as inflation, growth, market liquidity, and investor sentiment, instead of focusing on a single metric. The approach acknowledges that seemingly contradictory investment perspectives can both be valid amid market uncertainty, enabling simultaneous positioning for inflation and deflation, or for both economic growth and recession.

Practical Application: Quantum-inspired investors should build portfolios with exposure to multiple economic scenarios. They are likely to fare better than making concentrated bets on specific outcomes. It requires mapping how different assets perform in various environments and balancing exposures accordingly. The diversification encompasses different return-and-risk profiles that react differently to the same situation. This quantum perspective acknowledges that financial markets, like quantum systems, follow probability distributions rather than deterministic paths.

Startups pivoting business models

The Quantum State: Successful startups maintain superpositions of multiple potential business models. This allows them to preserve momentum and organizational cohesion even when initial strategies fail and adapt quickly to a new reality.

The Story: The evolution of Slack from a failed gaming company to a workplace communication giant represents one of the finest examples of startup pivoting. The process resembled a quantum-like superposition in business strategy. Founded in 2009 as Tiny Speck by Stewart Butterfield (previously co-founder of Flickr), the company initially focused on developing a massively multiplayer online game called "Glitch." After raising approximately $17 million in venture capital, Tiny Speck launched Glitch in 2011. Despite creative acclaim, the game failed to attract a sufficient user base and was shut down in 2012. This failure could have ended the company, but instead, it led to one of the most successful pivots in startup history.

During Glitch's development, the team created an internal communication tool to coordinate their distributed workforce. This tool was developed to have better collaboration within the team. This tool eventually became Slack.

What makes Slack's pivot particularly interesting is that the company didn't simply abandon one business for another. Instead, it maintained elements of both identities during the transition. Rather than concentrating solely on the product, they prioritized the benefits that customers are likely to value. This broader purpose helped the company maintain stability during the significant product change. It lets the team keep their culture and skills while fitting new market needs. Instead of creating a game that connected multiple players, they developed a system that facilitated multiparty conversations.

The pivot proved extraordinarily successful. Slack launched publicly in 2014 and reached a $1.12 billion valuation by October 2014, eight months after its public launch. This was one of the fastest journeys to "unicorn" status in venture capital history. By 2019, when the company went public via direct listing, it had around 12 million daily active users. In 2021, Salesforce acquired Slack for $27.7 billion.

There are other examples of similar quantum-like characteristics. X (erstwhile Twitter) began as Odeo, a podcast platform. They were rendered largely obsolete when Apple added podcasts to iTunes. Instagram started as Burbn, a location-based check-in app similar to Foursquare. They then pivoted to focus exclusively on photo sharing. In each case, the companies maintained aspects of their original identity while adapting to market feedback.

Successful pivots depend on staying flexible. Some companies view pivots as a binary decision. They drop their past fully, which hurts them. They typically struggle with losing team cohesion and momentum. Those that pivot successfully, like Slack, maintain continuity in purpose and culture while adapting their product strategy.

Quantum Finance Insights: The transformation of Slack is an example of strategic superposition in the world of startups. By holding onto several possible identities at once, a company can adapt quickly when market feedback narrows those options. Just as quantum particles exist in multiple states until measured, successful startups remain open to different directions until customer needs clarify the best path forward.

This concept is similar to the quantum observer effect, where observation alters the state of a system. Slack treated market feedback as more than just a verdict on their gaming venture – it became a guide for what the company could become. This perspective enabled them to adapt much more fluidly than competitors who only used feedback to confirm or reject their original vision. Successful pivots require founders to understand how their initial assumptions shape the way they interpret market signals and to use that awareness to remain flexible when reality challenges them.

Practical Application: Entrepreneurs can use quantum thinking to set broad missions for pivots. For example, shift investment from a failing app to a new tool while keeping the team united. This cuts risk and reallocates funds for better use. This means distinguishing between core purposes, which should remain stable, and specific implementation, which should adapt to market feedback. This quantum perspective acknowledges that startup success, like quantum systems, follows probabilistic rather than deterministic paths. They require constant adaptation based on market interaction.

Looking ahead

Financial superposition changes the way we deal with uncertain situations, allowing organizations to succeed by considering several suitable options at the same time until the market shows the best choice. From Bridgewater's All Weather Portfolio to Slack's masterful pivot, we've seen how embracing contradictions rather than resolving them creates robust, adaptive strategies. Yet even the most sophisticated superposition strategies face their ultimate test when market stability itself collapses. While maintaining multiple perspectives prepares us for various scenarios, some events transcend all predictions. These are the unforeseen events that disrupt our meticulously planned strategies. Next, we'll explore these quantum-like collapses, where entire financial systems leap from stability to chaos in ways that defy traditional risk models.

Chapter 9

The collapse of certainty: financial black swan events

"*Even the deepest root of the mighty oak does not know when soil shifts*," goes an old Quantaland proverb.

A major currency crisis triggered losses that snowballed in minutes, crushing hedge fund manager Omar's diversified fund. Crises overwhelmed his robust risk-management models. The speed and impact had never been projected by his models, even as an extreme event. This chapter explores quantum-like collapses where stable systems turn chaotic. Analysis of Black Swan events like oil price crashes will equip you with frameworks to distinguish predictable events from true black swans. It will also provide tools for adaptive crisis response systems that have a better chance of thriving in uncertainty.

Beneath the calm waters: how stability breeds catastrophe

Financial systems often exhibit a form of quantum-like metastability. They can maintain a prolonged period of apparent calm that masks underlying fragilities, only to undergo a sudden, catastrophic collapse when conditions exceed a critical threshold. These systems do not always deteriorate gradually. They can also experience rapid phase transitions, where a seemingly stable system transforms instantaneously into a highly unstable state.

Traditional financial analysis is built on the assumption that deterioration follows predictable, linear patterns. This creates a dangerous false assurance that problems will develop slowly, providing ample warning signals and time for intervention. Quantum finance, backed by overwhelming evidence from decades of crises, argues the contrary.

Many financial collapses are not the culmination of slow bleeding. A careful reconstruction of the events will reveal that they are discontinuous transitions. Systems can leap between fundamentally different states,

DOI: 10.4324/9781003619635-9

from stable to unstable, from liquid to illiquid, and from solvent to insolvent, all without any warning. Understanding these quantum-like collapse mechanisms is essential to explaining the persistent market mystery of why financial crises consistently manage to surprise even the most seasoned participants.

The architecture of fragility: hidden flaws in complex systems

Market fragility is not a weakness but a structural characteristic. It is a byproduct of the complex interdependence that drives efficiency during stable periods. These same connections create quantum-like vulnerabilities that remain invisible until stress reveals them, transforming latent links into channels of instantaneous contagion.

Hidden vulnerabilities in complex systems

Financial systems often develop their greatest weaknesses during periods of their own apparent success. Efficiency improvements and innovation in good times quietly breed dangerous blind spots that only a crisis can expose.

Quantaland Bank appeared to have a robust mortgage portfolio that was perfectly diversified across various regions and customer types. However, this geographic spread concealed the fact that lending standards had loosened simultaneously across all regions. During normal times, these mortgages existed in a quantum superposition, appearing both conservative and aggressive depending on the lens. The 2008 housing crash acted as a measurement, collapsing this superposition into a single, harsh reality, revealing that the "conservative" standards were, in fact, universally reckless.

The same principle applies to derivative markets, where risk appears widely distributed until a systemic event reveals that everyone is exposed to the same underlying shock. Quantaland Ports had diversified their disaster insurance across multiple companies. Hurricane Zeta revealed that all of Quantaland Ports' insurers were exposed to the same catastrophic event, albeit through different routes. Their smart diversification was actually a hidden concentration.

Algorithmic trading amplifies this problem. In calm markets, thousands of algorithms appear independent and diverse. Under stress, they often behave like a single herd, all executing the same orders simultaneously. What looks like healthy variety in good times demonstrates dangerous conformity at the precise moment when true diversification is needed most.

Quantum-like phase transitions in markets

Financial markets do not always decline gradually. They often undergo sudden, quantum-like phase transitions, leaping from stable to chaotic states in an instant.

Liquidity is the perfect example. One moment, markets function smoothly with tight bid-ask spreads. Next, a phase transition occurs. Buyers vanish, and liquidity evaporates without warning. Trading costs explode, and the market enters a new, unstable state.

Quantaland Trading experienced this during the March 2020 COVID crash. Their profitable strategies, honed over years in normal markets, became useless within days as market conditions underwent a quantum leap. The same pattern cripples credit markets, where lending can vanish overnight even for solvent borrowers. The same happens in currency markets, where stable correlations break down in a frantic flight to safety.

The VIX volatility index embodies this, often surging from calm sub-15 levels to panic levels above 30 in a single session. These are not adjustments; they are discontinuous jumps to new market realities where all the old rules cease to apply.

The illusion of stability during calm periods

Calm, low-volatility periods are often not states of stability but dangerous illusions. They are temporary equilibriums that breed vulnerabilities and encourage the very behaviors that make the eventual collapse more severe.

Falling volatility misleads risk systems into assuming safety, which in turn encourages increased leverage and higher concentrations of risk. Diversification benefits become mirages. When a crisis strikes, portfolios that appear well balanced in calm times become highly correlated, wiping out the protection investors counted on.

Before 2008, the Quantaland Pension Fund was caught in this trap. Their models interpreted steadily lowering volatility as proof of declining risk, leading them to load up on mortgage securities that seemed safe. This apparent stability hid the massive systemic fragility that exploded when the housing market collapsed.

This complacency infects everyone. Traders become accustomed to easy execution, borrowers to cheap credit, and regulators to quiet markets, which they often mistake as evidence that their oversight is working. The calm itself becomes the source of the coming fragility.

Cascade effects and systemic breakdown

Financial crises spread through cascade effects, in which a problem at one institution triggers immediate and simultaneous failures throughout the entire system due to hidden connections.

These are not gradual contagions but quantum-like collapses. The failure of Lehman Brothers didn't ripple out slowly. In fact, it instantly created losses and liquidity problems everywhere through its entangled network of counterparties.

Quantaland Credit Union experienced this when its sound financial position became irrelevant the moment its main correspondent bank failed. Access to payment systems and funding vanished instantly, creating a crisis through no fault of its own.

These cascades operate through multiple channels. Stressed institutions create funding cascades by simultaneously competing for scarce funding, leading to the complete disappearance of normal lending. Forced selling triggers margin calls, which in turn force more selling, leading to a downward spiral that ignores the fundamental values.

When fear spreads through pure psychology, triggering bank runs between institutions with no direct business links, you see confidence cascades in action. Similarly, emergency interventions meant to calm markets trigger regulatory cascades, which intensify the panic by interpreting them as signals of profound weakness.

These effects reveal finance's true quantum nature. Measuring (or revealing) a problem at one point can potentially instantly collapse the state of the entire entangled system.

Liquidity evaporation and market freezes

Liquidity exhibits one of the most extreme quantum-like behaviors in finance: the ability to vanish instantaneously without any intermediate state of decline.

Market makers and liquidity providers echo this behavior. Instead of withdrawing gradually, their warning bells ring simultaneously, causing them to all step back at once. This, instead of reducing liquidity, eliminates it entirely.

Funding markets also freeze in the same way. Institutions that could borrow easily suddenly find no willing lenders at any price. Commercial paper, interbank lending, and repo markets can all seize up at once.

During the March 2020 crash, Quantaland Securities saw its bond trading algorithms shut down across all sectors simultaneously, contributing to a market-wide liquidity crunch. During the crash, bid-ask spreads, which were previously pennies wide, exploded into dollars. This signified a significant shift into a new market state where conventional trading mechanisms simply collapsed.

When the entire network of private liquidity provision collapses at once, central banks become the only viable solution. They can step in as the lender of last resort.

Nonlinearity in finance

Financial systems are fundamentally nonlinear, exhibiting quantum-like behavior where minuscule inputs can trigger massively disproportionate effects. Understanding this nonlinearity is essential to explaining why crises consistently blindside even the most experienced market participants and policymakers.

Exponential versus linear response patterns

Traditional finance is built on an assumption of linearity, which believes that small causes produce small effects. Reality often is far more explosive. Financial systems are wired for exponential responses, where tiny triggers are amplified by feedback loops and threshold effects that lead to catastrophic outcomes.

Leverage is the purest example. A 20% price drop can wipe out 100% of the equity in a 5:1 leveraged position, transforming a manageable decline into a total loss. Quantaland Hedge Fund experienced this in their mortgage strategy. A seemingly small 2% increase in defaults triggered margin calls and forced asset sales. This resulted in further price declines, creating a death spiral that vaporized a major portion of the fund.

Network effects rapidly amplify problems, as each failing institution can trigger multiple counterparty failures simultaneously, not sequentially. Confidence operates on a principle where minor concerns can trigger withdrawals that create genuine liquidity problems, seemingly validating the original fear and accelerating the panic.

Algorithmic trading supercharges this, where a small trade can trigger a cascade of stop-loss and momentum orders, generating price movements that dwarf the initial catalyst. Information systems act as the ultimate accelerant, spreading negative news through social media and algorithmic processing faster than human judgment can provide context. This development turns minor concerns into full-blown crises almost instantly.

Threshold effects and tipping points

Financial systems are riddled with hidden thresholds. These are critical points where gradual deterioration triggers a sudden, discontinuous leap into a new and dangerous state.

Capital ratio thresholds are a regulatory example. A bank can operate smoothly until its capital ratio dips just below a regulatory minimum. This tiny breach instantly triggers mandatory supervisory action, forcing emergency capital raising on highly unfavorable terms. The consequences for Quantaland Bank after a minor breach during a downturn were catastrophic.

The consequence – fueled by panic sourcing of capital – exceeded the damage the small shortfall itself could have caused.

Credit-rating changes create similar quantum jumps. A one-notch downgrade, especially a fall from investment grade to junk, can trigger massive, forced selling by institutional investors. It also leads to covenant violations and a dramatic spike in funding costs, regardless of the actual change in credit quality.

Margin thresholds and liquidity covenants operate with the same brutality. They function as tripwires that, when breached, trigger automatic, nonnegotiable responses. Consequent forced selling or the immediate acceleration of debt repayment can instantly transform a manageable situation into an existential crisis.

Feedback loops and self-reinforcing cycles

Financial systems are uniquely susceptible to destructive feedback loops, where an initial problem triggers a response that amplifies the original issue, creating a self-reinforcing cycle of decline.

The classic bank run is a great example of confidence loops. Fear leads to withdrawals, resulting in liquidity shortfalls that reinforce the fear and prompt additional withdrawals. Similarly, the credit loop starts with banks tightening lending in a downturn. This action weakens the economy, leading to more defaults and forcing further credit tightening.

A perceived rise in risk forces an institution to pay higher rates for funding, triggering a funding loop. The higher rates weaken their financial position, which forces them to pay even higher rates. Quantaland Investment Fund was caught in asset-liability loops. Market losses triggered investor redemptions, forcing asset sales in a falling market. These sales further drove prices lower, which amplified losses and triggered further redemptions, creating an accelerating downward spiral.

These loops are the engines of crisis, turning small problems into major collapses through quantum-like amplification.

Network effects in financial contagion

Modern financial networks are designed for efficiency, but they are also perfectly engineered for quantum-like contagion, allowing local problems to transform into global crises almost instantly.

Hub institutions such as major banks, CCPs, and payment systems act as super-spreaders. A problem at a single hub, Quantaland Clearing Corporation, instantly triggered liquidity demands across its entire network of members. This engulfed even those without direct links to the original issue.

Correlation amplification occurs through these networks. Different paths in the network may expose seemingly diversified institutions to the same underlying risk. A web of guarantees, derivatives, and interbank lending can create hidden concentrations that only become visible when the entire network is stressed.

Information cascades now spread at algorithmic speed. Negative news about one institution can trigger panic about connected entities through social media and automated news processing. This spreads more quickly than human interventionists can assess the actual strength of the connections.

Payment-system contagion also represents a pure network effect. An operational failure at one major bank can disrupt the flow of payments for thousands of other institutions that rely on that shared infrastructure. This event can potentially paralyze an entire economic region.

The mathematics of disproportionate impact

In addition to being a theoretical feature, the nonlinearity of finance is a mathematical reality that manifests itself in a variety of business features.

Consider leverage for one. A 10% decline in the underlying assets creates a 100% loss on a 10:1 leveraged position. This simple multiplicative effect is why Quantaland Hedge Fund's 15% market loss translated into a 150% fund wipeout. Similarly, the probabilities of systemic failure are not linear; they compound. If 14 independent systems each have a 5% chance of failure, the probability that at least one fails is given by $1 - (0.95)^{14} \approx 51.24\%$. A seemingly small individual risk aggregates into a more likely-than-not system-wide threat.

The probability of contagion doesn't increase linearly with the network density, which is the number of connections in a network. It increases exponentially. A densely connected system is disproportionately more fragile than a sparse one. Similarly, the impact of extreme events is so disproportionate. Even very low-probability outcomes must be respected, as their consequences are nonlinear and potentially catastrophic.

Beyond prediction: building resilience for the unknowable

Preparing for Black Swan events requires a fundamental philosophical shift. Since these crises lie outside all prediction frameworks, the goal cannot be to predict them. Instead, quantum-inspired preparation focuses on building adaptive capacity and strategic optionality. This builds the ability to respond effectively regardless of which specific scenario emerges.

Robustness versus resilience

Traditional risk management pursues robustness, trying to prevent disruptions through strong defenses and conservative practices. Quantum finance champions resilience. It is the capacity to adapt and recover when disruptions inevitably occur. Resilience provides superior protection because it doesn't require predicting specific threats.

Building systems that don't break versus systems that adapt

Robust systems resist shocks but may shatter under unprecedented force. Resilient systems bend, absorb, and reconfigure. Quantum uncertainty suggests that resilience often provides superior protection because it doesn't require predicting specific threats.

Quantaland Bank learned this distinction over the years. Their post-2008 robust capital buffers excelled at absorbing credit losses but proved useless against COVID-19's operational challenges. They discovered that capital strength alone couldn't address the need for rapid adaptation to remote work, digital service delivery, and supply-chain disruptions.

This resilience principle applies across all operations. Operational resilience focuses on maintaining business continuity through adaptation capabilities rather than preventing all disruptions. Robust systems use redundancy to prevent failures but cannot guarantee technology resilience. Resilient systems, on the other hand, use modular architecture and rapid response capabilities to adapt when failures occur.

To build supply-chain resilience, Quantaland Manufacturing shifted from efficiency-focused chains to resilience-focused approaches. They maintained multiple supplier relationships and geographic diversity. Similarly, to achieve financial resilience, one needs to maintain multiple funding sources and covenant flexibility rather than optimizing only for the lowest-cost financing.

Antifragile strategies that benefit from chaos

One of the most sophisticated quantum approaches is antifragility, involving positioning to benefit from volatility and uncertainty rather than simply surviving them. Optionality approaches build antifragile traits by keeping several possible responses that can be adjusted as situations change. Quantaland Technology exemplifies this through their venture capital approach to innovation, maintaining small investments in multiple emerging technologies. These can be rapidly scaled if market conditions prove favorable.

Barbell strategies combine extremely safe investments with small high-risk positions that can deliver enormous payoffs from positive surprises. This strategy effectively acknowledges that predicting extreme events is impossible.

Volatility harvesting strategies benefit directly from market uncertainty, using instruments that gain value from volatility increases regardless of market direction. Similarly, network effect strategies build businesses that strengthen as networks grow, creating positive feedback loops where increased adoption boosts gains.

Learning systems approach antifragility through capabilities that improve with experience and stress, and become stronger through challenges.

Redundancy versus flexibility in system design

This critical design choice separates conventional and quantum preparation. Redundancy provides backup capabilities for known failures, while flexibility provides adaptation capabilities for unknown challenges.

Financial systems maintain excess capital and liquidity buffers for known stress scenarios, but these may not respond well to unpredictable challenges. Operational approaches emphasize adaptable processes and capabilities rather than fixed procedures designed for specific scenarios. In both cases, the focus is on flexibility.

Another example is technology systems that use modular architecture and cloud computing for rapid scaling and reconfiguration. This is how they avoid building fixed infrastructure useful for specific capacity requirements. Workforce flexibility demands maintaining contractor relationships and cross-training capabilities that provide a rapid response to changing business requirements.

Capital structure flexibility maintains unused debt capacity and diverse funding relationships rather than optimizing only for current cost minimization. Quantaland Insurance balances this well, maintaining excess capital for known catastrophe exposures while preserving flexibility to adjust strategies when new risks emerge.

The quantum superposition of preparedness strategies

Quantum-inspired preparedness means maintaining a superposition of strategies simultaneously. It avoids single strategies based on predicted crisis types, recognizing that actual crises often differ from predicted scenarios.

Scenario superposition maintains preparedness for multiple crisis types, such as financial, operational, and external shocks, at once rather than optimizing for single scenarios. Likewise, response capability superposition maintains multiple options where deployment is based on actual crisis characteristics. This encompasses financial resources, operational flexibility, and adaptive decision-making processes.

Resource-allocation superposition maintains multiple channels and message frameworks that are activated based on actual crisis characteristics. This helps preserve the capability to address various stakeholder concerns.

This approach avoids the fatal trap of optimizing for the wrong crisis. Instead of predetermined plans, it builds the deep capability to respond, adapt, and profit regardless of what form the uncertainty takes.

The quantum insight is clear: certainty collapses under measurement. Your preparation must therefore remain in a superposition of multiple readiness states, multiple response capabilities, and multiple strategic options until the crisis itself forces the collapse into specific action.

Crisis management for a quantum world

Traditional crisis management plans often assume that crises will follow predictable patterns. They further assume that planned responses will remain appropriate when actual crises occur. Quantum-inspired crisis management recognizes that actual crises often differ from planned scenarios and require adaptive response capabilities.

Pre-positioning for unknown unknowns

Effective crisis preparation must account for events that lie entirely outside planning scenarios. Quantum-inspired approaches focus on building generalized capabilities and decision frameworks rather than predicting specific events.

Response capability building develops organizational tools such as decision-making frameworks, communication systems, and resource mobilization processes that adapt to various crisis types. Detailed plans of Quantaland Corporation for operational disruptions proved useless against pandemic lockdowns that eliminated customer demand entirely. They needed improvised tools their specific plans hadn't envisioned.

Stakeholder relationship building creates crisis response options through strong ties with customers, suppliers, employees, and regulators. During actual crises, these relationships often prove more valuable than any contingency plan.

Resource optionality maintains flexible tools that can be deployed based on actual crisis characteristics. Cash reserves provide optionality for various responses, while dedicated insurance only addresses specific risks.

Decision authority structures establish clear crisis command channels that function when normal processes disintegrate. Learning system development, on their part, creates organizations that learn and adapt during crises rather than executing predetermined plans.

Dynamic crisis response frameworks

Static crisis plans become obsolete the moment a crisis diverges from expectations. Dynamic response systems provide structured adaptability while maintaining organizational coherence.

Situation assessment tools provide systematic approaches to understanding unprecedented conditions when predetermined plans are useless. This can be supported by resource reallocation systems, enabling rapid redirection of resources toward crisis response, suspending normal operations to concentrate on immediate threats.

Communication protocols provide systematic approaches to stakeholder communication when normal channels fail. Quantaland Services developed such systems after supply-chain disruptions exposed their plan's inadequacies. Their new framework emphasizes rapid assessment and flexible response rather than executing predetermined scripts.

Escalation procedures provide tools for expanding response when initial actions prove inadequate, preventing persistence with ineffective strategies. Similarly, recovery planning integration ensures that crisis actions consider long-term needs and avoid solutions that create longer-term problems.

Communication strategies during quantum uncertainty

Crisis communication during Black Swan events requires fundamentally different approaches than corporate messaging during stable periods. Acknowledging uncertainty means admitting real doubts instead of projecting false confidence. Quantaland Financial discovered this with their candid and truthful disclosures of their plans of action during the crisis. Stakeholders appreciated honest uncertainty more than competitors' confident predictions that repeatedly proved wrong.

Multiple perspective strategies maintain simultaneous messaging for different stakeholder concerns and time horizons. This avoids a single message-driven strategy that misses diverse audience needs. Real-time adaptation adjusts messaging based on evolving conditions and stakeholder responses, abandoning static plans that become obsolete.

Stakeholder-specific communication recognizes that customers, employees, regulators, and investors need different information and approaches during crises. Trust maintenance strategies focus on preserving confidence at times when stakeholders make decisions with incomplete information. This is achieved through transparency, avoiding over-promising and under-delivering.

Decision-making under extreme time pressure

Black Swan events demand rapid decisions with incomplete information when normal processes collapse. Rapid decision systems provide structured approaches that work under time pressure. Quantaland Bank developed such systems after facing funding crises that demanded immediate action, enabling critical decisions in hours instead of weeks. The bank achieved this by integrating the capabilities of information systems with data analytics.

Authority delegation structures ensure decision-makers can act without extensive consultation, preventing fatal delays when rapid response is critical. Information prioritization systems help focus on essential information while avoiding paralysis from overload. At the same time, decision reversibility assessment identifies which choices can be modified later versus those creating irreversible commitments.

Stakeholder impact assessment provides rapid tools for understanding how decisions affect different groups. This enables the anticipation of reactions and the resultant adjustment of implementation tools.

Learning systems that adapt during crises

Quantum uncertainty requires that effective crisis management involve continuous learning and adaptation during the crisis itself. Real-time learning tools capture information about crisis conditions and response effectiveness to improve ongoing management. Quantaland Financial implemented such tools to track customer behavior changes and operational challenges during major disruptions. Further, these tools continuously adapted their response.

Experimentation methods provide systematic approaches to testing new responses when traditional approaches fail. This enables the discovery of effective solutions through controlled experimentation rather than random improvisation. Feedback integration ensures crisis response teams receive rapid feedback about effectiveness and unintended consequences that require modifications.

Knowledge-sharing tools capture lessons learned and share them across organizational units facing similar challenges. Adaptation protocols provide procedures for modifying responses when initial approaches fail or when crisis conditions change significantly.

Understanding the quantum nature of financial Black Swan events transforms crisis preparation from specific scenario planning to building adaptive systems that respond effectively to truly unprecedented challenges.

Meanwhile, far from Quantaland

The 2020 pandemic's economic impact

The Quantum State: Global economies existed in a state of apparent stability. It collapsed almost instantaneously when the pandemic triggered simultaneous supply and demand shocks across interconnected systems.

The Story: The COVID-19 pandemic's impact on the global economy in 2020 represents one of the most dramatic examples of certainty collapse. A localized health crisis in Wuhan, China, rapidly evolved into a global economic cataclysm. The impact of the event fundamentally challenged established economic assumptions and policy frameworks.

The sequence of events unfolded with unprecedented speed. On December 31, 2019, Chinese authorities reported cases of pneumonia in Wuhan to the World Health Organization (WHO). By January 23, 2020, Wuhan entered lockdown. On March 11, the WHO declared COVID-19 a global pandemic. By mid-March, approximately 100 countries had implemented full or partial lockdowns, bringing global economic activity to a sudden halt.

The market reaction was immediate and severe. Between February 19 and March 23, 2020, the S&P 500 fell 34%. This was the fastest decline of that magnitude in history. Oil futures briefly traded at negative prices in April as storage capacity reached limits. Global air travel decreased by approximately 60% in 2020, while sectors such as hospitality, retail, and entertainment experienced revenue declines of 50–90%.

What made the pandemic particularly disruptive was the unprecedented scale and speed of simultaneous supply and demand shocks. While dual shocks have occurred historically, conventional economic models have struggled to address the rapid, global nature of this collapse. There was no textbook solution to prevent what was happening.

The policy response was equally unprecedented. The Federal Reserve cut interest rates to near zero and implemented open-ended quantitative easing. This expanded their balance sheet by approximately $3 trillion in just three months. The U.S. Congress passed the $2.2 trillion CARES Act in March 2020, which was the largest economic stimulus package in American history. Similar measures were implemented globally, with the IMF estimating that worldwide fiscal support reached $16 trillion by March 2021.

Perhaps most revealing was how the pandemic exposed the fragility of systems previously considered robust. Just-in-time supply chains, optimized for efficiency rather than resilience, collapsed under pandemic pressures. Commercial real estate valuations, dependent on stable office occupancy, faced fundamental reassessment as remote work became normal. Entities such as restaurants, airlines, and others whose business models depended on physical presence required complete reinvention. The response was also one of the swiftest and largest policy interventions in history. However, some industries such as e-commerce and communication technology thrived, showing asymmetrical impacts.

The Quantum Finance Lesson: The economic turmoil experienced during the pandemic vividly illustrated how certainty within financial systems can abruptly collapse. Structures that once appeared steadfast can rapidly dissolve the moment their underlying assumptions are overturned. This sudden transition mirrors the behavior of quantum wave functions, which shift from broad probability distributions to defined states when observed.

This phenomenon is akin to quantum decoherence, where a system loses its predictable behavior when confronted by external disturbances. The pandemic served as a catalyst for economic decoherence, forcing theoretical economic models to confront real-world biological disruptions. These challenges were beyond the scope of the models' design.

The crisis made clear that economic stability is sustained not only by the objective assessment of fundamentals but also by collective belief in shared narratives. Yet, as the pandemic showed, this confidence can dissipate with surprising speed when faced with new and contradictory information.

Practical Application: Risk managers should incorporate quantum thinking by stress-testing systems against failures to fundamental assumptions. They are usually restricted to stress-testing quantitative variations. This requires identifying the core premises upon which business models depend and developing contingency plans for scenarios where those premises become invalid. There are times when the models may be theoretically sound but only up to a scale. Stress tests should take into account the scale beyond which the models fail.

We have seen in earlier case studies that some firms, like Walmart, adapted dynamically by reconfiguring supply chains; most systems lacked the resilience to handle such a rapid collapse.

As finance professionals, we need to acknowledge that financial stability, like quantum states, depends on probability distributions that can collapse almost instantaneously when critical variables change.

Oil price collapse (2014–2016)

The Quantum State: Oil markets appeared stable under OPEC's price management, but this equilibrium was eroding due to rising U.S. shale production. When geopolitical and technological shifts fully unraveled OPEC's control, a new supply-demand paradigm emerged.

The Story: The 2014–2016 oil price collapse represents a classic example of how seemingly stable market structures can disintegrate when underlying assumptions change. After nearly four years of relative stability with Brent crude averaging around $110 per barrel, oil prices plummeted over 70% between June 2014 and January 2016. The prices bottomed out at approximately $28 per barrel, the lowest level since 2003.

A confluence of factors fundamentally altered the structure of the oil market, leading to this dramatic decline. The U.S. shale revolution had quietly transformed global supply dynamics. U.S. production nearly doubled from 5.5 million barrels per day in 2010 to 9.3 million by 2015. This made America the world's largest oil producer and reduced its import dependence. Simultaneously, global demand growth slowed as China's economy decelerated and energy efficiency improved worldwide.

The collapse of certainty came on November 27, 2014, when OPEC, led by Saudi Arabia, made the shocking decision. They decided not to cut production despite plunging prices. This represented a fundamental shift in Saudi strategy. They moved away from price defense to market share protection.

This decision directly challenged the market's core assumption that OPEC would act as the global oil market's central bank. They were expected to adjust supply to maintain price stability. The impact was immediate and severe. Oil prices fell 10% the day after the announcement and continued to decline throughout 2015 as a global supply glut developed. U.S. oil companies slashed capital expenditures by approximately 50% between 2014 and 2016, while over 100 North American oil and gas producers filed for bankruptcy.

The geopolitical impacts were equally significant. Russia, Venezuela, Nigeria, and other oil-dependent economies experienced severe recessions and currency crises. Some Gulf states like the UAE and Qatar weathered the storm better due to sovereign wealth buffers. Saudi Arabia itself began using their foreign reserves at an unprecedented rate, eventually implementing austerity measures. Reserves fell from $732 B (2014) to $555 B (2015). Value-added tax was introduced in 2018, and subsidies were reduced. Then came the groundbreaking decision to diversify its economy away from oil dependence.

Perhaps most telling was how the price collapse challenged fundamental valuation models across the energy sector. The assumption that oil prices would inevitably return to $100+ levels was challenged. This assumption, based on peak oil theory and rising global demand, was proven wrong as the hidden vulnerabilities came out in the open.

By 2016, a new equilibrium began emerging. U.S. shale production finally declined in response to lower prices, while global demand gradually increased. OPEC eventually returned to active market management, partnering with Russia in December 2016 to form OPEC+. They implemented production cuts that helped stabilize prices around $50–60 per barrel. This was roughly half their pre-collapse levels. These prices were below pre-2014 levels, confirming a structural shift, not a full return to old norms.

The Quantum Finance Lesson: This market collapse echoes the concept of quantum phase transitions, where systems can abruptly shift from one state to another when critical thresholds are met. The oil market, once in a stable phase under OPEC's control, saw that equilibrium overturned by the rise of shale technology and a strategic pivot from Saudi Arabia. As a result, the market transitioned from managed price stability to a new regime of competitive price discovery. This echoed the behavior of quantum systems that change dramatically when their core properties are altered.

This transformation reflects the quantum principle of breaking spontaneous symmetry, where a system moves between different organizational states depending on shifting conditions. The symmetry of the oil market was rooted in OPEC's price-setting authority. This order fractured as technological advances elevated U.S. shale producers as key marginal suppliers. In this way, the oil market revealed complexity and adaptability far beyond the simplified models that often directed investment strategies.

Practical Application: Commodity strategists should identify potential phase transitions where market structure could fundamentally change instead of focusing on merely forecasting price movements within existing paradigms. This means monitoring technological innovations, strategic shifts among key players, break-even prices of marginal producers, and divergences between theoretical models and market behavior.

This quantum perspective acknowledges that commodity markets, like quantum systems, can exist in multiple organizational states, with transitions between them triggered by critical threshold events.

Swiss franc unpegging (2015)

The Quantum State: The Swiss franc existed in a policy-enforced superposition. It was pegged at 1.20 CHF/EUR, masking its true market value. When the Swiss National Bank (SNB) ceased interventions, the currency collapsed into its "eigenstate," revealing the underlying economic reality. To clarify, an eigenstate is the single, definite state revealed upon measurement; for example, the cat is either alive or dead.

The Story: On January 15, 2015, the SNB shocked global financial markets by abruptly abandoning its three-year-old currency peg. This pegging had maintained a minimum exchange rate of 1.20 Swiss francs per euro. This unexpected policy reversal triggered one of the most violent currency moves in modern financial history. The Swiss franc appreciated by approximately 30% against the euro within minutes of the announcement.

The events leading to this "franc shock" began in September 2011. The SNB established the currency floor to combat deflationary pressures and protect Swiss exporters from a rapidly appreciating franc during the Eurozone debt crisis. Maintaining this peg required the SNB to purchase foreign currencies in virtually unlimited quantities. This caused its balance sheet to expand dramatically from 46% of Swiss GDP in 2011 to over 80% by early 2015.

The catalyst for abandoning the peg came from the ECB's impending quantitative easing program, which was expected to weaken the euro significantly. Continuing to maintain the floor peg would have required even more massive currency interventions that the SNB deemed unsustainable. In these circumstances, the SNB did not find enforcing and maintaining the minimum exchange rate for the Swiss franc against the euro to be justified any longer.

What made this policy change particularly devastating was its complete surprise. The common sentiment, aided by the actions of the SNB, was that the peg would remain a cornerstone of Swiss monetary policy. This created a false sense of certainty that amplified the shock when the policy changed suddenly.

The market impact was immediate and dramatic. Within minutes, the EUR/CHF exchange rate plunged from 1.20 to as low as 0.85 before partially recovering to around 0.97 by day's end. The Swiss stock market (SMI) fell 8.7% in its largest one-day decline in 25 years. Companies with significant Swiss operations saw their shares plummet. For example, the watchmaker Swatch Group fell 16%, while Holcim (now LafargeHolcim) dropped 15%. Beyond financial markets, the economy as a whole also experienced severe disruption. Swiss exporters faced an instant competitive disadvantage as their goods became approximately 15–20% more expensive in foreign currency terms. Tourism operators reported immediate cancellations as Switzerland suddenly became much more expensive for foreign visitors.

The collateral damage extended globally. Many brokers had no liquidity buffers for such a gap move. Several FX brokers, including UK-based Alpari and New Zealand's Excel Markets, declared insolvency after clients incurred losses they couldn't cover. FXCM, then the largest retail forex broker in the United States, required a $300 million bailout from Leucadia National Corporation after losing $225 million. Perhaps most dramatically, global macro hedge fund Everest Capital closed its flagship $830 million fund after suffering crippling losses from the franc's move. Other funds like Discovery Capital also suffered heavy losses.

Quantum Finance Insights: The Swiss franc unpegging serves as a vivid example of how certainty in currency markets can suddenly collapse. Artificial stability, sustained by policy intervention, may vanish instantly when those interventions are withdrawn. This mirrors the quantum principle, where probabilities become definite outcomes upon observation. Similarly, the currency's value shifted from a controlled range to its real market level.

Much like a quantum system under constant watch, the Swiss franc's rate was artificially held in place by the SNB's actions. When these interventions ended, the currency quickly returned to its natural value, experiencing a rapid and dramatic adjustment that exposed previously hidden market forces. With the removal of external controls, these forces surfaced powerfully and abruptly.

Practical Application: Risk managers should recognize that policy-maintained market equilibria represent artificial states. These states can collapse instantaneously when policies change. This calls for stress-testing positions against extreme scenarios even when current policies appear stable. This also calls for maintaining hedges against sudden regime shifts in market structure. They should also recognize that many hedges fail in gap moves, which may require liquidity diversification.

This quantum perspective acknowledges that market stability, like quantum states, often depends on continuous observation and intervention. This, in turn, creates the potential for dramatic phase transitions when that intervention ceases.

Looking ahead

Black Swan events reveal the ultimate limits of prediction-based financial planning, demonstrating how even the most sophisticated risk models crumble when fundamental assumptions collapse. The COVID-19 pandemic, oil price crashes, and currency shocks examined by us show that true preparedness requires building adaptive systems rather than predicting specific outcomes. Yet within this apparent chaos lies opportunity for those who understand how to navigate extreme uncertainty. While black swans destroy conventional strategies, they also create the conditions where quantum-inspired approaches can truly shine. Our next exploration reveals how individual investors can transform this understanding into practical decision-making frameworks that thrive on uncertainty rather than being paralyzed by it.

Chapter 10

Quantum decision-making for investors

"*The river meanders till rocks lead it to the sea,*" goes an old Quantaland proverb.

Retiree Wei relied on a growth-focused fund. She saw her savings shrink in the 2022 tech crash, as she could not move away from her fixed strategies, which had served her very well over time. This chapter transforms investment decision-making from prediction-based to probability-driven. It will show you how ARK Invest and Yale's endowment address shifting probabilities. These stories will teach you to embrace uncertainty, implement superposition thinking, and develop adaptive systems that thrive in dynamic markets.

The quantum investor's mindset: from prediction to probability

Quantum investing demands a fundamental shift in mental models. It moves beyond traditional approaches that seek certainty through analysis, instead embracing market uncertainty as an inherent and exploitable characteristic. This mindset doesn't just manage risk; it creates unique opportunities for those who understand how to navigate probabilistic landscapes effectively.

This transition represents more than a tactical adjustment. It requires reinventing the investment process itself. The quantum investor abandons the futile question, "What will happen?" and replaces it with the more powerful inquiry, "What could happen, and how do we position ourselves to benefit from multiple potential outcomes?"

Uncertainty as fuel: turning risk into opportunity

Quantum investing is based on a key change in thinking that sees uncertainty not as something to fix but as a natural part of the market that

DOI: 10.4324/9781003619635-10

can be used to your advantage. This transformation turns uncertainty from an obstacle into your most valuable resource for achieving superior performance.

Moving from prediction to probability thinking

Traditional investing seeks to predict future conditions and position accordingly. Quantum investing treats markets as probability distributions, not fixed paths, requiring strategies that perform well across multiple potential scenarios.

Quantaland Asset Management exemplifies this shift. They abandoned single-point forecasts for probability ranges: modeling a 25% chance of negative returns, 50% for 0–15%, and 25% for returns above 15%. This enables portfolio construction that thrives across scenarios rather than betting on a single outcome.

Point estimates create dangerous illusions of precision. Predicting that Alpha Corporation's earnings will grow by 12% and Beta Corporation's by 8% suggests a certainty that doesn't exist. Quantaland Pension Fund avoids this by weighing scenarios. They may consider a 40% probability of low inflation, 35% of moderate, and 25% of high inflation, and then build a portfolio that reflects these weights rather than betting on a single forecast.

Uncertainty as a source of opportunity rather than risk

Quantum investing recognizes that uncertainty creates option value and trading opportunities that certainty eliminates. Markets reward those who navigate doubt well but punish those who demand certainty before acting.

Volatility harvesting strategies profit from mispriced options during high-doubt periods. Quantaland Hedge Fund systematically sells option premiums when implied volatility is high, betting that subsequent realized volatility will be lower.

Uncertainty premiums appear across asset classes in which investors demand higher returns for uncertain outcomes. Credit spreads widen during volatility, creating opportunities for willing investors. Emerging markets often trade at discounts due to uncertainty premiums rather than fundamental value differences.

Quantaland Real Estate Investment Trust profits by acquiring properties during market stress when temporary liquidity premiums allow purchases below long-term value. The highest returns often come from investing when volatility peaks and consensus is most negative, which is precisely the point when many investors retreat.

Quantum superposition in investment analysis

Investment analysis exhibits quantum-like behavior where securities appear simultaneously attractive and unattractive depending on the analytical framework. Quantum investors maintain exposure so they can benefit from multiple perspectives.

Valuation superposition occurs when DCF suggests undervaluation while P/E ratios suggest overvaluation. Quantaland Equity Research tracks multiple metrics without forcing a single price target, reflecting a range of possibilities.

Strategy superposition maintains simultaneous exposure to many key areas. The strategy includes growth and value styles, quality and momentum factors, exposure to multiple sectors instead of predicting rotations, international diversification rather than focusing on specific countries, and other elements.

This approach recognizes that different strategies dominate during different periods and that consistently predicting these shifts is nearly impossible.

The effect of measurement in investment decisions

The act of analysis in investment research creates quantum-like measurement effects that influence the behavior under study. Understanding these impacts is crucial for accurate assessment. The impact of research on prices occurs when widespread analysis affects market behavior. Positive reports can trigger buying pressure that eliminates the very undervaluation they identified.

Quantaland Securities observed that their positive reports caused short-term price increases in the highlighted stocks, reducing valuation attractiveness. They now adjust their exposure to account for these measurement effects.

Index inclusion effects further demonstrate how measurement changes behavior. Stocks joining major indices exhibit increased correlation, and valuation is influenced regardless of fundamentals.

ESG measurements create particularly powerful observer effects. Quantaland Sustainable Investing has documented how including companies in their portfolio actually shapes those companies' sustainability practices. Investment analysis ends up being an active force influencing strategy rather than a passive observation.

The living portfolio: algorithms that learn and adapt

Quantum investing leverages adaptive algorithms that respond to evolving market conditions rather than following static, rule-based approaches.

These systems embody quantum thinking by maintaining flexibility across multiple potential market states and adjusting their behavior as market measurements reveal the prevailing conditions.

Machine learning applications in portfolio management

ML provides the computational engine for quantum investing, offering tools that identify patterns and adapt strategies based on real-time market evolution. These applications move beyond traditional optimization to create truly adaptive investment systems.

Pattern-recognition algorithms identify recurring market patterns without predetermined hypotheses. Quantaland Technology employs ML algorithms that analyze thousands of market variables to detect possible trends by predicting a range of short-term movements. These systems continuously adapt as market microstructure evolves, maintaining effectiveness where traditional technical analysis becomes obsolete.

Regime detection algorithms identify when markets transition between different behavioral states, enabling strategies that adapt to changing conditions. Correlation modeling uses ML to track evolving relationships between securities, capturing new diversification opportunities as market interdependencies shift.

Quantaland's risk managers use ML systems that monitor thousands of risk indicators to identify emerging threats before they materialize. These systems evolve continuously as new risk factors emerge and existing risks change in importance.

Trade execution optimization represents another critical application, where ML algorithms adapt to changing market microstructure. Instead of following static rules, they attempt to improve execution quality and reduce costs.

Real-time strategy adjustment based on market conditions

Adaptive algorithms enable real-time strategy adjustments that respond to changing market conditions faster than the speed of human decision-making. These systems maintain multiple strategic possibilities, allowing evolved market states to select the optimal approaches.

Volatility-based adjustments modify portfolio risk exposure using current market fluctuations rather than assuming a static risk tolerance. Quantaland Portfolio Services uses algorithms that change risk exposure every day by looking at volatility measurements, changes in correlations, and liquidity factors. They maintain target risk levels while adapting to evolving market conditions.

Application possibilities are varied. For example, momentum/mean reversion switching recognizes markets that alternate between trending and oscillating patterns. Liquidity-based adjustments modify trading strategies based on current market depth and execution costs. Sector allocation systems adapt continuously rather than following predetermined rotation schedules.

The Quantaland Strategy Board employs algorithms that monitor relative sector performance, economic indicators, and technical patterns to identify optimal allocations in real time. This reduces the chance of missing actual market transitions.

Behavioral adaptation in algorithmic trading

Algorithmic trading must account for interactions with other algorithms and human traders, creating adaptive systems that respond to changing participant behavior.

Market microstructure adaptation requires algorithms to adjust to the evolving market structure. Quantaland Trading adapts execution algorithms based on current bid-ask spreads, market depth, and volume patterns, optimizing for actual rather than assumed market conditions.

Competitive algorithm adaptation acknowledges that trading systems must account for other algorithms pursuing similar strategies. This requires continuous adjustment to maintain effectiveness.

Order flow analysis enables algorithms to identify changing patterns that affect price discovery and execution quality. Quantaland Arbitrage Fund employs algorithms that monitor price relationships across equity, bond, and derivative markets to identify temporary mispricing, adapting continuously as market efficiency changes.

Time-of-day adjustments recognize that market behavior varies across trading sessions, requiring algorithms that adjust based on the current time and expected participant activity levels.

Quantum-inspired optimization techniques

Optimization techniques inspired by quantum mechanics provide superior methods for portfolio construction by maintaining multiple potential solutions simultaneously.

Quantaland Investments uses advanced algorithms that keep several portfolio setups ready for different market situations. This strategy allows for swift adjustments instead of needing a full overhaul when things change.

These techniques include methods that consider entanglement and evolving interdependencies among portfolio components. Another technique is uncertainty propagation, which maintains probability distributions around input assumptions instead of using static parameters.

Another technique involves multi-scenario optimization that creates portfolios designed to perform well across multiple economic scenarios. Multi-objective approaches that balance competing objectives such as return maximization, risk minimization, and liquidity management simultaneously can be useful, too.

Feedback loops in adaptive investment systems

Adaptive investment systems create feedback loops where performance influences future behavior. This requires careful design to avoid destructive cycles while capturing beneficial adaptation effects.

Performance feedback systems enable algorithms to learn from both successful and unsuccessful trades. Quantaland Learning Systems employs algorithms that analyze trade results to identify successful patterns while avoiding overreaction to random outcomes.

Risk feedback mechanisms adjust risk tolerance and position sizing based on portfolio performance and market conditions. Exposures are increased after successful periods and reduced after losses or stress periods.

Correlation monitoring tracks how portfolio changes affect asset relationships and adjusts future composition accordingly. Quantaland Multi-Market Investments uses feedback systems to watch how different investment strategies work together, adjusting their allocations based on what they see happening instead of assuming each strategy works alone.

Adjustment systems for market impact modify trading strategies based on observed effects from previous trades. These systems learn to minimize impact while maintaining execution quality through an adaptive approach.

Navigation, not maps: dynamic allocation for changing terrain

Dynamic asset allocation represents the practical application of quantum thinking to portfolio management as well as to fund allocation in corporate finance. Under this approach, asset allocations adapt continuously to changing market conditions rather than following static strategic allocations based on long-term assumptions.

Market regime detection and response

Financial markets show varied behaviors during growth, recession, and crisis periods, so they require different asset allocation approaches. Effective tactical strategies depend on identifying current market regimes and adjusting allocations accordingly.

Economic cycle identification uses multiple indicators to determine current phases that influence optimal asset allocation. Let us look at a few similar regime changes. Volatility regime analysis recognizes that markets alternate between low and high-volatility periods, which require different risk-management strategies.

Correlation regime monitoring tracks how asset class correlations change in different states. Crisis periods often show increased correlations, so they reduce diversification benefits and require allocation adjustments.

Quantaland Pension Fund employs regime detection models that analyze employment data, inflation trends, central bank policies, and market indicators. They adjust strategies based on regime identification rather than following static strategic plans. Quantaland Tactical Allocation employs models that identify current market momentum characteristics and emphasizes different strategies accordingly. They maintain exposure using both momentum and mean reversion approaches, so they emphasize whichever matches current behavior.

Central bank policy regimes recognize that monetary policy changes create different market environments, affecting optimal strategies.

Dynamic rebalancing strategies

Dynamic rebalancing adapts portfolio maintenance to changing market conditions, so it avoids calendar-based or threshold-based rules that ignore context. These strategies optimize rebalancing timing and magnitude based on current conditions.

Volatility-adjusted rebalancing modifies frequency and magnitude based on current market volatility. Momentum-aware rebalancing accounts for asset momentum when determining timing and magnitude. Strong momentum periods justify delayed rebalancing, allowing exhaustion to indicate optimal timing. Quantaland Portfolio Services employs fluctuation-adjusted rebalancing, which increases frequency during volatile periods when allocations drift rapidly. They reduce frequency during calm periods to minimize transaction costs.

Cost-benefit analysis weighs transaction costs against expected benefits, accounting for current liquidity and expected return differences. Liquidity-sensitive rebalancing adjusts methods based on current market liquidity conditions, affecting transaction costs and market impact. Tax concerns also play an important role. Tax-efficiency-focused portfolio managers in Quantaland coordinate rebalancing decisions with tax optimization approaches. Their focus is to minimize tax impact while maintaining target allocations. They harvest tax losses while optimizing asset location across taxable and tax-efficient accounts.

Corporate finance applications of tactical allocation

Tactical allocation principles, in addition to portfolio management, apply to corporate finance decisions such as cash management, capital structure, and investment timing. They require dynamic strategies rather than static policies.

Corporate cash management represents tactical allocation between cash, short-term investments, and operating assets. It adapts to changing business conditions and market opportunities. This is why Quantaland Manufacturing employs tactical cash management that adjusts its allocation based on seasonal working capital needs, investment opportunities, and market factors. They maintain the required liquidity and optimize returns on excess cash.

There are many applications of tactical allocation in corporate finance. Capital structure optimization adapts debt-equity ratios, looking at changing market conditions. The parameters include interest rates, credit spreads, and equity valuations. Working capital management uses tactical strategies to optimize inventory, receivables, and payables. It further adapts to changing business conditions, seasonal patterns, and customer behaviors.

Quantaland Services adapts their dividend policy based on cash-flow stability, investment opportunities, and market factors. They avoid static payout ratios that become suboptimal during changing business factors.

Investment timing decisions also apply tactical principles to capital expenditure timing, acquisition approaches, and research spending. This helps entities benefit from market timing while maintaining their strategic objectives.

Volatility-based allocation adjustments

Fluctuating conditions significantly influence optimal asset allocation because they affect both risk and return characteristics across asset classes. Adaptive allocation systems adjust portfolio risk exposure based on current and expected market shifts.

Risk parity adjustments modify asset allocations to maintain consistent risk contributions across portfolio components as fluctuations change investment profiles. Quantaland Finance employs fluctuation-based allocation adjustments that maintain target risk levels. They adapt to changing market fluctuations and automatically adjust allocations to ensure consistent risk exposures across varying conditions.

On the other hand, fluctuation-forecasting models predict near-term changes that influence optimal allocation decisions. They help identify when conditions may shift, so they create opportunities for preemptive adjustments.

Smart volatility managers employ strategies that harvest fluctuation premiums through systematic options trading. They adjust portfolio exposure by dynamically reallocating assets. Correlation analysis of market swings recognizes that correlation patterns change during different regimes and affect diversification benefits. Therefore, allocation adjustments are required to maintain portfolio risk characteristics.

ESG integration in tactical decisions

Sustainability factors increasingly influence tactical allocation decisions. They affect risk and return characteristics across asset classes and individual securities. Sustainability momentum analysis identifies companies and sectors with improving characteristics, so they may outperform when these factors influence market performance.

Regulatory sustainability trends affect asset allocation as government policies influence business operations and investment flows. This happens through features such as climate change regulations, social responsibility requirements, and governance standards. ESG risk assessment identifies how environmental and social risks affect allocation decisions across regions and sectors.

Quantaland Green Fund employs impact momentum analysis to identify allocation opportunities among companies with sustainability-improving practices. They favor companies with strong impact momentum and avoid those with deteriorating characteristics. They adjust allocations to optimize financial returns and measurable social and environmental impact across varying conditions.

Inclusive capitalism influences allocation toward companies adopting broader-focused business models. This allows these companies to perform better during times that prioritize value creation, thereby transcending the focus on maximizing shareholder profits.

The tactical edge: responsive positioning in real time

Tactical asset allocation implements quantum thinking by dynamically adjusting portfolio weights in response to evolving market conditions. At the same time, strategic long-term goals are not overlooked. This approach recognizes that optimal allocations exist as probability distributions that change as market conditions evolve.

Risk-management techniques

Quantum-inspired risk management acknowledges that risk exists in dynamic probability distributions rather than as static parameters. Effective risk management must therefore adapt continuously to changing market conditions and correlation patterns, moving beyond fixed approaches that become obsolete.

Dynamic hedging strategies

Dynamic hedging adapts hedge ratios and timing to changing market conditions, avoiding static positions that become ineffective when relationships change. These strategies optimize hedging effectiveness while minimizing costs across various asset classes.

Currency hedging optimization adjusts hedge ratios based on evolving currency volatility, correlation patterns, and forward-premium characteristics. Interest-rate approaches adapt duration exposure based on changing rate volatility and yield-curve dynamics. Equity techniques modify ratios and selections based on market volatility and correlation patterns.

Quantaland Energy employs dynamic commodity hedging that adjusts ratios based on production schedules, price volatility, and inter-commodity correlation patterns. They adapt to changing market conditions rather than following static policies, recognizing that traditional hedging relationships often break down during crisis periods.

Cross-hedging methods use related instruments when direct hedging is unavailable. You have to continuously adapt to changing correlation relationships between the instruments and underlying exposures.

Stop-loss evolution and quantum thresholds

Traditional stop-loss strategies often fail during quantum market transitions when prices leap beyond predetermined levels. Adaptive approaches evolve with market conditions and account for quantum-like price movements.

Volatility-adjusted levels modify stop-loss triggers based on current market volatility rather than on fixed percentages. Quantaland Trading employs volatility-adjusted systems that widen stop-loss levels during high-volatility periods while tightening during calm conditions, maintaining consistent risk exposure across different market environments.

Time-based stop evolution recognizes that optimal levels change as positions age and market conditions evolve. Technical pattern analysis identifies optimal levels based on support and resistance rather than on arbitrary percentages and adapts to the current market structure.

Quantaland Portfolio Management, when implementing strategies, employs correlation-adjusted systems that consider portfolio-level effects and market liquidity conditions. This approach avoids the pitfall of treating individual positions in isolation.

Corporate hedging applications

Where companies face multiple uncertainty sources, they require integrated approaches rather than isolated strategies. Corporate hedging, in such cases, should apply quantum-inspired risk management to various facets of their business operations.

Operational hedging combines business strategy and financial hedging to establish natural hedges within operations. Quantaland Manufacturing achieves operational hedging by diversifying its production facilities and supply chains across different geographic locations. This provides natural protection against currency fluctuations, regulatory changes, and economic disruptions.

Integrated risk management coordinates hedging across multiple exposures, including currency, commodity prices, interest rate, and credit risk. This approach acknowledges the complex interactions between these exposures.

Cash-flow hedging focuses on protecting stability rather than balance sheet values by using instruments that reduce volatility. It is immaterial that these hedges may not qualify for hedge accounting.

Competitive hedging considers how risk-management strategies affect competitive position when competitors have different exposures or approaches. This approach accounts for industry-wide practices and dynamics.

Stress testing and scenario-based risk management

Quantum-inspired stress testing recognizes that extreme scenarios often differ from historical experience. This characteristic calls for probabilistic approaches that account for multiple potential conditions. Scenario generation creates multiple extreme cases reflecting different potential market disruptions. This moves beyond a single worst-case scenario that may not capture actual conditions.

Dynamic stress testing adapts scenarios based on changing market conditions and emerging risks rather than using static cases that become obsolete. Current geopolitical tensions, regulatory changes, and market structure evolution require tests that use updated extremes.

Quantaland Investment Management employs Monte Carlo stress testing that generates thousands of potential scenarios based on historical patterns and hypothetical conditions. This model evaluates the portfolio's performance across multiple potential outcomes.

Correlation stress testing, another useful approach, examines portfolio behavior when asset correlations increase dramatically during stress periods. These increased correlations often eliminate diversification benefits when they are most needed. Liquidity stress testing evaluates performance when market liquidity evaporates and transaction costs rise significantly.

Quantaland Asset Management employs comprehensive stress testing that includes correlation breakdown scenarios, liquidity crisis conditions, and tail-risk events exceeding normal assumptions.

Behavioral risk management

Behavioral risk management addresses how human psychology creates significant risks that traditional financial models often ignore. Quantum-inspired methods consider how psychological factors influence investment decisions and market outcomes.

Cognitive error identification helps investors recognize systematic thinking errors, including overconfidence, anchoring, and loss aversion. All these create predictable patterns of suboptimal decision-making. Emotional management addresses how fear and greed influence investment choices during different market conditions, often leading to buying high during euphoric periods and selling low during panic.

Quantaland Investment Committee employs structured processes that require independent analysis from multiple team members before reaching decisions. This process prevents groupthink that often leads to poor choices.

Bias calibration helps investors develop accurate assessments of their predictive abilities and knowledge limitations. Overconfident investors often take excessive risks and make frequent trading decisions based on illusions of control or misinterpreted market signals. They end up reducing returns through transaction costs and poor timing.

Meanwhile, far from Quantaland

ARK Invest's strategy during the 2022 tech sell-off

The Quantum State: ARK maintained investment convictions while acknowledging shifting probability distributions. They adjusted position sizes and time frames rather than abandoning their core innovations during market turbulence.

The Story: ARK Investment Management's decisions during the 2022 technology sector downturn represent an instructive case study in quantum decision-making under extreme market pressure. Founded by Cathie Wood in 2014, ARK gained prominence for its focus on disruptive innovation. They were involved in areas such as genomics, fintech, autonomous technology, and next-generation internet. After a spectacular performance during 2020–2021, when its flagship ARK Innovation ETF (ARKK) gained 152% in 2020 alone, the firm faced a dramatic reversal in 2022. Rising interest rates and inflation concerns hammered growth stocks, causing this decline.

The magnitude of this reversal was extraordinary. After peaking at $159.70 in February 2021, ARKK fell to a low of $33.00 in December 2022, clocking a 79% decline that erased most of its pandemic-era gains. Assets under management across ARK's ETFs plummeted from approximately

$50 billion to $15 billion. This crash also correlated tightly with the rise of 10-year treasury yields from 1.5% to 4.2%. During these declines, traditional media outlets and value investors were strongly critical.

What makes ARK's response noteworthy is its nuanced combination of conviction and adaptation. Rather than abandoning its innovation thesis or doubling down blindly, they adjusted its approach in quantum-like fashion. They maintained core positions while acknowledging shifting probability distributions around time frames and valuations.

This adaptability was evident in ARK's trading patterns. The firm used market volatility to consolidate its portfolios, reducing holdings from approximately 60 stocks in early 2021 to around 30–35 by late 2022. This consolidation represented a quantum-like adjustment of probability weighting rather than a binary pivot in strategy. Simultaneously, ARK adapted its time-horizon messaging. They maintained conviction in the technological transformations driving their investments. But at the same time, they acknowledged that higher interest rates and economic uncertainty had extended the timeline for some innovations to reach commercialization.

They admitted the difficulty in predicting a timeline for innovation but were sure that those innovations would eventually generate extremely high returns. ARK's trading revealed quantum-like non-commutativity where the order of operations mattered. Selling liquid winners (TSLA) to buy illiquid losers (TDOC) amplified drawdowns, much like quantum measurements disturbing observed systems.

Perhaps most revealing was how ARK communicated with different stakeholder groups during the downturn. With existing investors, the firm emphasized patience and historical context. They emphasized that previous innovation cycles had experienced similar periods of volatility before delivering exponential returns. At the same time, with prospective investors, ARK highlighted the attractive entry points created by the sell-off, arguing that innovation was "on sale." This dual messaging reflected quantum-like complementarity. They acknowledged that both short-term pain and long-term opportunity could simultaneously be valid perspectives.

The firm also demonstrated adaptive transparency during this period. Rather than hiding from criticism, they faced it strongly. They had enhanced their public presence, published research defending ARK's positions, hosted more frequent investor webinars, and actively engaged with skeptics on social media. They acknowledged valid critiques and simultaneously defended their innovation thesis.

By early 2023, this quantum approach showed signs of validation as ARKK rebounded approximately 40% from its December 2022 low. While still far below its peak valuations, the recovery suggested ARK's adaptable-conviction strategy had positioned it to benefit from renewed interest in innovation themes.

The Quantum Finance Lesson: ARK's actions during the downturn serve as a prime example of quantum-inspired decision-making in investment management. By embracing probabilistic thinking instead of relying solely on deterministic views, ARK was able to adapt to changing market conditions without losing sight of its core beliefs. Much like quantum systems, which are defined by distributions of probabilities rather than fixed states, ARK viewed its innovation thesis as a collection of potential outcomes, each with its own probability. The volatility of the 2022 sell-off acted as a measurement event, shifting short-term expectations, such as profit timelines, while still preserving the long-term potential for significant innovation gains.

In quantum theory, the wave function evolves smoothly under the Schrödinger equation rather than leaping abruptly between states. Similarly, ARK adapted its investment approach gradually, integrating new information while holding onto its foundational principles. The firm recognized that while individual innovation outcomes remained uncertain and subject to change, the fundamental drivers of technological advancement endured even amid market upheaval.

Practical Application: Investors should separate investment objects into probabilistic components rather than treating them as monolithic entities. This means adjusting position sizes and time horizons based on evolving evidence while maintaining conviction in well-researched secular trends. This quantum perspective acknowledges that investment success, like quantum systems, involves navigating probability distributions. It does not make binary, all-or-nothing bets on specific outcomes.

David Swensen's Yale endowment model

The Quantum State: Swensen's approach acknowledges fundamental economic uncertainty by maintaining simultaneous exposures to multiple asset classes and strategies. This allows continuous adaptation without requiring forecasting precision.

The Story: David Swensen's Yale endowment model represents one of the most influential applications of quantum-like thinking to portfolio construction. As chief investment officer of Yale University's endowment from 1985 until his death in 2021, Swensen transformed institutional investing. He grew the endowment from $1 billion to over $31 billion by 2020. His strategy explicitly recognized the fundamental uncertainty of financial markets and built portfolios that could thrive across multiple potential economic scenarios.

The genesis of the Yale Model came from Swensen's insight that traditional stock-bond portfolios inadequately addressed true market uncertainty. He emphasized the need to diversify broadly across asset classes with low correlations, which could prove useful in protecting against the unknowable future. This led to a radical shift from conventional 60/40 stock-bond mixes. Instead, Swensen allocated heavily to alternative assets such as private equity, real estate, and absolute return strategies.

The Yale Model typically allocates across six main categories, namely absolute return, domestic equity, fixed income, foreign equity, private equity, and real assets. Let's take a brief look at what these categories represent. Absolute return focuses on event-driven and value-driven strategies for consistent gains; domestic equity refers to U.S. stocks for growth; fixed income focuses on bonds for stability; foreign equity means international stocks for diversification; private equity is comprised of venture capital and buyouts for high returns; and, finally, real assets include real estate, timber, and energy for inflation protection. Swensen's allocation wasn't just diversification. It resembled a wave function composed of complementary assets. When one "collapsed," like equities in 2008, others, like private equity and real assets, maintained the portfolio's phase coherence.

The Yale Model constructs a quantum-entangled portfolio. Going beyond superposition, it creates an entangled system where asset classes interact differently under varying economic measurements such as inflation, growth shocks, and so on. Each component performs well in specific environments, while the overall portfolio maintains resilience no matter which scenario emerges.

What sets the Yale Model apart from traditional diversification is its focus on illiquid alternatives and active management. Swensen recognized that correlations between assets change over time. They depend on economic regimes. This is a quantum-like observation that different market contexts reveal different relationship patterns.

The model proved its worth during major crises. In 2008, when the S&P 500 dropped 37%, Yale's endowment fell by 24.6%. It outperformed peers thanks to real assets and absolute return strategies. Again, during the COVID-19 shock in early 2020, Yale's diversified approach limited losses to 2.7% for the fiscal year ending June 2020. Stock volatility hit many traditional portfolios much harder during this time.

This resilience comes from embracing uncertainty rather than predicting outcomes. Swensen avoided market timing. He focused on long-term commitments to high-conviction managers in alternative spaces. He was open to accepting uncertainty and prepared for multiple futures, not focusing on one.

The Yale Model has influenced endowments worldwide. Institutions like Harvard and Princeton have adopted similar strategies. They have shifted toward alternatives for better risk-adjusted returns.

The Quantum Finance Lesson: The Yale Model approaches financial markets much like a quantum field, illustrating that embracing uncertainty can result in more resilient investment outcomes than relying on precise predictions. In this analogy, major economic regimes, like periods of inflation or recession, act as disturbances within the field. Alternative assets serve as individual probes, each interacting with the market environment in unique ways. The act of rebalancing plays a role similar to quantum measurement, determining the portfolio's position among possible outcomes. Just as quantum physics replaced deterministic certainty with probability distributions, the Yale Model favors broad, balanced exposures over singular economic predictions.

This idea echoes the quantum principle of superposition, where a system remains in multiple states until it is observed. Similarly, the Yale portfolio is positioned for a range of economic possibilities at any given time. It is designed to withstand shocks like sudden data releases or shifts in policy that force portfolios to react to new dominant regimes. By accepting uncertainty as a fundamental aspect of investing, the Yale Model prepares for a spectrum of future scenarios rather than relying on a single, specific forecast.

Practical Application: Investors should incorporate quantum thinking by constructing portfolios that acknowledge economic uncertainty and do not rely on precise forecasts. This means mapping how different assets respond to various regimes and maintaining balanced exposures.

Diversification is not driven merely by statistical inference. It requires thinking beyond traditional assets. This quantum perspective also acknowledges that markets follow probability distributions. It believes, and not without reason, that resilience would hold over precision in uncertain environments.

Value investing versus growth investing

The Quantum State: Like quantum systems, the best investors exist in a superposition of value and growth states. Macro shocks (e.g., inflation) act as "measurements," temporarily collapsing the portfolio toward one style, but the underlying strategy remains hybrid.

The Story: The historical tension between value and growth investing philosophies represents one of finance's most enduring debates. Value investing, pioneered by Benjamin Graham and championed by Warren Buffett, focuses on purchasing securities trading below their intrinsic value. Growth investing, advocated by investors like Philip Fisher and T. Rowe Price, emphasizes companies with above-average growth potential regardless of their current valuation metrics.

For decades, these approaches were often presented as opposing philosophies requiring exclusive commitment. However, the investment landscape of the 21st century has revealed that the most adaptable investors maintain quantum-like superpositions of both approaches. They adjust their relative emphasis based on market conditions while incorporating elements of each philosophy.

This complementary perspective became particularly evident during the extraordinary market cycles of 2010–2022. From 2010 to 2020, growth strategies dramatically outperformed value, with the Russell 1000 Growth Index returning 17.2% annually versus 10.5% for its value counterpart. This outperformance peaked during the pandemic, with growth stocks benefiting from accelerated digital adoption while value sectors such as energy and financials struggled.

Yet this relationship reversed sharply in 2022, when rising interest rates and inflation concerns triggered a dramatic style rotation. The Russell 1000 Value Index outperformed its Growth counterpart by 21.6 percentage points. This was one of the largest annual performance differentials in modern market history post-1970s, when the Russell index was introduced.

What distinguishes quantum investors during these transitions is their ability to maintain elements of both philosophies simultaneously. Warren Buffett, traditionally viewed as the archetypal value investor, demonstrated this flexibility. He made significant investments in technology companies like Apple (becoming its largest individual shareholder by 2018) while continuing to apply valuation discipline to these growth-oriented positions.

Similarly, growth-oriented investors like Baillie Gifford's Tom Slater adapted their approaches during the 2022 downturn without abandoning their fundamental philosophy. They recognized the opportunities that exist in changing times.

Perhaps the most instructive example of quantum investing comes from those who explicitly reject the value-growth dichotomy entirely. Let us take the case of Bill Miller, whose Legg Mason Value Trust famously outperformed the S&P 500 for 15 consecutive years. He countered the popular view that value investing means paying low multiples. He redefined value investing as buying companies for less than their worth. This perspective acknowledges that growth itself is a component of value! This perspective recognizes that investment philosophies, which may seem opposed, can actually complement each other within a unified approach.

The evolution of quantitative investing further illustrates this convergence. Factor-based strategies increasingly incorporate both value and growth signals rather than treating them as mutually exclusive. Quant funds mimic quantum entanglement: value and growth factors are non-separable. Research by *AQR Capital Management*, known for factor investing, shows their correlation isn't static. In reality, they exist in a probabilistic relationship where one factor's collapse, like a value rally, reveals the other's state. They found that combining value and momentum factors, which are often associated with growth investing, produced superior risk-adjusted returns compared to either approach in isolation. Since value premiums are largely unpredictable, it makes sense to hold both value and growth stocks in the portfolio.

The Quantum Finance Lesson: The relationship between value and growth investing serves as a vivid illustration of quantum complementarity in financial strategy. Rather than viewing these approaches as mutually exclusive, this perspective underscores how each offers a unique yet harmonious lens on market reality.

This idea reflects the quantum principle of complementarity, where different methods reveal distinct aspects of a system that may seem at odds. But in reality, they complement one another. In physics, light can appear either as a particle or a wave, depending on how it is observed. Similarly, investors may evaluate companies using either growth criteria or value criteria, depending on the prevailing market conditions. Ultimately, value and growth embody complementary and not conflicting ways to measure corporate worth.

Practical Application: You should incorporate quantum thinking by maintaining elements of both value and growth philosophies in your analytical frameworks. This is a safer approach to facing quantum uncertainty than committing exclusively to either approach. This means you need to develop valuation models that explicitly consider growth prospects while applying valuation discipline to growth-oriented positions.

Smart investors will be open to both growth and value approaches. This quantum perspective acknowledges that investment success, like quantum reality, requires embracing complementary frameworks rather than forcing false dichotomies.

Looking ahead

The quantum investor's journey from prediction to probability represents more than a strategic shift. It is a fundamental reimagining of how markets actually work. From ARK Invest's adaptive resilience during the 2022 tech crash to Yale's enduring success with quantum-like diversification, we've seen how embracing uncertainty creates superior long-term outcomes. Yet individual investing is only part of the quantum finance revolution. The same principles that transform personal portfolio management can also revolutionize corporate finance itself. While investors navigate probability distributions in markets, corporate leaders face quantum decisions about capital allocation, financing structures, and strategic optionality that determine their companies' survival and success. Next, we'll explore how CFOs and corporate leaders can apply these quantum principles to build more adaptive, resilient business operations.

Chapter 11

Quantum corporate finance

"*The bee who knows many gardens survives many storms*," goes an old Quantaland proverb.

CFO Nia committed to a massive debt-financed expansion when a demand drop turned it into a costly burden. Her plan, based on assumptions of stability, couldn't adapt to changing conditions. This chapter fuses corporate finance with quantum principles, showing how to use real options and flexible budgets. The chapter examines dynamic capital allocation, financing, and distribution policies. It concludes by describing frameworks for stakeholder communication and ESG integration, which are likely to thrive in quantum market conditions.

The quantum CFO: leading with flexibility and optionality

Corporate finance leadership in quantum markets demands a fundamentally different skill set from traditional financial management. The quantum corporate finance leader must move beyond optimizing static financial structures and build adaptive, responsive systems that maintain strategic optionality in the face of constant change.

This evolution from optimization to adaptation represents more than a mere tactical shift. It requires a complete reimagining of the finance function's purpose. It looks forward to a shift from the traditional role of finance as a cost center focused solely on efficiency to being a strategic enabler empowered with flexibility and dynamic responsiveness.

Allocation in superposition: keeping options open

Capital allocation represents one of the most critical areas where quantum thinking transforms corporate performance. Traditional approaches seek to optimize returns through careful analysis and deployment to the highest-value opportunities. In contrast, quantum capital allocation

DOI: 10.4324/9781003619635-11

maintains optionality and adapts resource deployment as market conditions reveal new opportunities and threats.

Real options in capital-allocation decisions

Capital allocation mirrors quantum superposition by holding options for future outcomes. Investments enable firms to scale, pivot, or exit as market conditions dictate the optimal path.

Quantaland Technology's R&D approach exemplifies this. Instead of betting on one massive project, they fund multiple smaller research initiatives. Each project functions as a call option on future market opportunities, where success means scaling up while failure limits losses to the initial investment.

Platform investments showcase this thinking perfectly. When Quantaland Manufacturing invests in flexible production systems, they purchase options to manufacture different products, serve new markets, and adapt to changing customer needs without major new capital commitments.

Geographic expansion operates through staged market entry. Quantaland Retail tests new markets with small pilot stores. They effectively function as options for full market entry if conditions prove favorable, preserving capital without overcommitting to uncertain opportunities.

Technology acquisitions and strategic alliances follow similar patterns. Companies make small strategic investments or partnerships that create options for full acquisitions later. This creates valuable optionality even when immediate benefits appear limited.

Modular investment strategies and scalability

Modular investments create quantum-like flexibility, enabling companies to scale up or down based on changing conditions. This is in contrast to making irreversible commitments to fixed investment levels.

Production capacity benefits from modular approaches that adjust to demand fluctuations. Quantaland Chemical uses modular production systems where additional capacity modules can be added quickly during demand surges and idled during downturns. This modular production structure offers a level of adaptability that fixed systems cannot provide.

IT infrastructure demonstrates modularity through cloud computing and scalable software choices. These platforms allow capacity to expand or contract based on business requirements while avoiding overbuilding.

Similarly, workforce scalability creates human capital flexibility through contractor relationships and flexible employment arrangements. This maintains core competencies while adapting to changing needs.

Supply-chain modularity maintains multiple supplier relationships across different regions. Quantaland Electronics keeps relationships with suppliers in various geographic areas, enabling quick sourcing shifts when trade disputes or disasters disrupt specific suppliers.

Cross-functional resource reallocation

Quantum corporate finance enables rapid resource shifts across business functions as markets evolve, overcoming the silos that traditional budgeting creates. Dynamic budget systems let companies shift resources between functions outside annual budget cycles. Quantaland Services employs monthly resource-allocation meetings that redirect resources from stable areas to emerging opportunities or problem areas. This maintains budget discipline while enabling tactical flexibility.

Project portfolio rebalancing shifts resources between projects based on changing success probabilities. Subject to availability and the abandonment option, this approach is more useful than funding projects for their entire planned durations regardless of performance.

Shared services optimization creates resource pools supporting multiple business units. Centralized functions like IT and finance can provide scalable support that adapts to changing needs.

Crisis resource mobilization involves the extreme reallocation of resources. During an epidemic, Quantaland Healthcare rapidly shifts its budget from elective services to epidemic response. This strategy allows it to maintain operations while supporting community health needs.

Quantum superposition in project evaluation

Project evaluation recognizes that projects exist in superposition across multiple potential outcomes until market conditions determine actual results. Scenario-based valuation evaluates projects across multiple market conditions rather than optimizing for a single expected scenario. Quantaland Energy evaluates renewable energy projects across various regulatory scenarios, commodity prices, and technology costs. They prefer accounting for scenario ranges rather than relying on point estimates.

Probability-weighted returns account for uncertainty by weighing different outcomes based on their likelihood. This provides more realistic evaluations that acknowledge uncertainty. Similarly, option value recognition includes expansion, abandonment, and timing options in project valuation. This approach acknowledges that many projects create strategic value beyond immediate returns.

Embedded sequential decision modeling recognizes that projects involve multiple decision points where companies can change paths based on evolving conditions. This model prevents the treatment of projects as single-commitment decisions.

Portfolio effects consideration recognizes that project values depend on how they interact with existing operations. Quantaland Manufacturing evaluates how projects developing new products use existing capabilities and complement current products.

Dynamic budget allocation systems

Traditional annual budgeting creates rigidity that prevents adaptation to changing conditions. Dynamic budget systems enable frequent resource shifts while maintaining financial discipline and strategic focus.

Rolling forecast systems replace annual budgets with continuous forecasting. Quantaland Corporation employs rolling 12-month forecasts updated monthly based on actual performance and market conditions. In this process, they maintain strategic planning focus while accommodating tactical adjustments.

Zero-based budget elements require periodic justification of ongoing expenses rather than assuming historical spending patterns remain optimal. This review enables resource reallocation from less effective activities to emerging opportunities. Similarly, trigger-based reallocation systems automatically adjust resources when performance or market indicators reach predetermined thresholds. This promotes rapid response without management intervention for routine adjustments.

Performance-based allocation adjusts resources based on actual results rather than maintaining fixed allocations. Quantaland Technology allocates development resources based on project milestone achievement and market feedback. Monitoring achievements permits rewarding success while enabling resource movement from struggling projects.

Strategic flexibility versus operational efficiency

Quantum corporate finance involves finding a balance between being adaptable to changes and running operations efficiently. This approach helps to keep options open for sudden market changes while also improving short-term results.

Capacity utilization demonstrates this balance. High utilization reduces costs but often eliminates the ability to respond to higher demands or production disruptions. An unserved customer is likely to change suppliers. Quantaland Manufacturing maintains 85% capacity utilization rather than maximizing efficiency, preserving flexibility to handle unexpected events without major capital investments.

Organizational structure decisions balance specialization against cross-functional capabilities. Highly specialized organizations achieve operational excellence but struggle when market conditions require different capabilities.

Technology investments balance standardization benefits against flexibility needs. Standardized systems provide cost advantages, while flexible systems enable adaptation to changing requirements.

Inventory management demonstrates efficiency-flexibility trade-offs. Lean inventory reduces costs but increases vulnerability to supply disruptions. Quantaland Retail adjusts their strategy during supply disruptions from lean optimization toward strategic stockpiling. This helps them maintain product availability while competitors face stockouts.

The responsive organization: building financial agility

Financial agility enables companies to adapt their financial operations quickly in response to changing market conditions, competitive threats, and emerging opportunities. This capability becomes particularly crucial during quantum market transitions, where traditional financial planning approaches become inadequate and potentially dangerous. Organizations that master financial agility can respond to disruptions while maintaining strategic focus and operational stability.

Organizational structures for rapid decision-making

Traditional corporate hierarchies often create decision bottlenecks that prevent a rapid response to changing conditions. Quantum corporate finance requires agile structures that enable quick decision-making while maintaining appropriate governance and risk management. This organizational approach resembles the quantum principle of superposition, maintaining multiple potential decision paths until market conditions demand specific actions.

Decentralized authority structures push financial decision-making to business units and operational managers, who can respond quickly to local conditions. Quantaland Systems exemplifies this approach by delegating significant financial authority to regional managers for pricing, inventory, and staffing decisions. This structure enables rapid market response while maintaining overall financial discipline through clear guidelines and reporting.

Several structural approaches support agile decision-making. Cross-functional teams integrate finance professionals with operations, marketing, and strategy teams, enabling decisions to be made faster than sequential departmental reviews. Executive decision committees create senior management groups with authority to make major financial decisions quickly, often convening virtually to decide within hours rather than waiting for scheduled board meetings.

Crisis management structures provide predefined authority and communication protocols for emergency situations, bypassing normal approval processes when speed becomes critical. Digital workflow systems automate routine financial decisions while escalating exceptional situations to appropriate authority levels.

Quantaland Financial Services employs automated approval systems for routine credit decisions while routing complex cases to specialist underwriters. This approach permits processing standard applications within minutes rather than queuing them behind more complex proposals. This approach significantly improves their responsiveness to customer needs.

Information systems enabling real-time financial adaptation

Real-time financial information systems form the technological backbone of financial agility. They provide current data about business performance and market conditions that enable rapid decision-making. Traditional periodic reporting cycles become inadequate when situations change quickly and require an immediate response.

Modern organizations employ several critical systems. One of the most common systems is a real-time dashboard that monitors KPIs across the organization. Quantaland Retail, for example, tracks daily sales, inventory turnover, and margins across all locations, enabling rapid response to shifts in customer behavior.

Predictive analytics that forecasts near-term performance and identifies emerging risks or opportunities provides early warning capabilities for preemptive action. This is a marked improvement over postmortem decision-making. Market intelligence integration that combines internal operational data with external market information promotes comprehensive situational awareness.

Automated reporting systems that generate routine reports while automatically flagging exceptional conditions requiring management attention support an agile decision-making process. Mobile access platforms are a great enabler. They enable decision-makers to access critical information and exercise approval rights from anywhere.

Quantaland Technology provides key metrics and approval authority to senior managers directly on their mobile devices. Availability of key data ensures that financial decisions can be made promptly, regardless of physical location.

Crisis response capabilities and financial resilience

Financial resilience enables companies to maintain operations and adapt strategies during crisis situations when normal business assumptions no longer apply. Building this resilience requires planning for unknown scenarios rather than simply optimizing for expected conditions.

Liquidity management systems ensure adequate cash availability when normal funding sources may become unavailable. Quantaland Energy maintains committed credit facilities that exceed normal operating requirements specifically to ensure liquidity during commodity price volatility or operational disruptions.

Effective crisis preparedness includes several key elements of operation. Scenario planning capabilities enable rapid strategy adjustment when market conditions change dramatically. This influences decision frameworks and resource reallocation procedures. Business continuity planning ensures that essential operations can continue during various disruption scenarios, including natural disasters, cyberattacks, and pandemics.

Recovery planning frameworks provide structured approaches for returning to normal operations after crisis conditions resolve. Another important component is the stakeholder communication systems. This system enables rapid communication with employees, customers, suppliers, and investors during crises. Awareness of the situation is often the major differentiator between a successful and failed recovery attempt.

Quantaland Insurance developed comprehensive business continuity plans after experiencing hurricane disruptions. These plans enabled continued customer service while maintaining appropriate financial controls during emergency conditions. To achieve this, they used procedures specifically designed for crisis situations rather than for normal operations.

Regulatory compliance in dynamic environments

Regulatory compliance becomes significantly more complex during dynamic business environments. Companies must adapt operations while complying with regulations that were often designed for more stable conditions. Quantum corporate finance requires compliance approaches that acknowledge and adapt to changing requirements while maintaining regulatory integrity.

Adaptive compliance systems continuously monitor regulatory requirements and adjust procedures based on changing business operations and regulatory interpretations. Quantaland Bank employs such systems, adjusting risk-management procedures based on evolving business activities and regulatory guidance.

Several strategies help manage compliance in dynamic environments. One of the strategies involves proactive regulatory engagement through regular communication with regulators about business changes and compliance approaches. Comprehensive documentation systems are very useful for demonstrating compliance in a dynamic environment. They maintain records of business decisions and compliance rationales to show good-faith efforts.

Another great technique is compliance flexibility planning that identifies areas where regulations provide interpretive flexibility. Similarly, cross-jurisdictional coordination systems that manage compliance requirements across multiple regulatory environments may be a preferred option for multijurisdiction entities.

Quantaland Global maintains compliance coordination systems that ensure consistent risk-management standards across all operating jurisdictions. At the same time, they adapt to local regulatory requirements to ensure jurisdiction-specific compliance.

Cultural change management for financial agility

Building financial agility requires significant cultural shifts that encourage adaptation, experimentation, and rapid decision-making rather than risk aversion and bureaucratic processes. These cultural changes often represent the most challenging aspect of implementing quantum corporate finance principles.

The development of risk tolerance involves encouraging appropriate risk-taking that enables business adaptation and growth. Traditional corporate cultures often emphasize risk avoidance to such an extent that it prevents companies from responding effectively to changing market conditions. Quantaland Innovation addresses this through programs that reward calculated experimentation and learning from failure.

Key cultural elements supporting financial agility require a multidimensional approach. You need to emphasize decision speed and prioritize timely action over exhaustive analysis when market conditions require a rapid response. Similarly, one has to value flexibility and recognize adaptation as more valuable than consistency when market conditions demand evolution. You will also need to support innovation by providing resources and encouragement for developing new approaches to business challenges.

Quantaland Designs maintains innovation labs where employees can develop new business ideas without the constraints of normal operational requirements. Such sandboxes create space for creative solutions to emerge outside of traditional processes.

Adapting capital structures to market changes

Capital structure optimization in quantum markets requires dynamic approaches that adapt to changing conditions rather than a static optimization based on historical relationships. Optimal capital structures exist in a superposition across multiple market scenarios until specific conditions determine the most appropriate financial configuration.

Debt versus equity financing

The choice between debt and equity financing involves complex trade-offs influenced by market conditions, business requirements, and strategic objectives. Quantum corporate finance maintains financing flexibility that can adapt to changing circumstances rather than committing to rigid financing approaches.

Dynamic capital structure optimization

Traditional capital structure theory assumes stable market conditions and predictable cash flows that enable optimization for single scenarios. Quantum capital structure recognizes that optimal configurations must evolve with changing market conditions and business requirements.

Quantum financing decisions respond to current debt and equity market conditions rather than assuming constant capital costs. Quantaland Development times its debt issuance based on interest-rate cycles and credit spread conditions. Similarly, their equity issuance timing responds to stock market valuations and investor sentiment.

Flexibility preservation maintains financing capacity for future opportunities, rather than optimizing for current capital costs. Companies that maximize current leverage may eliminate their financing flexibility and suffer when market conditions change unexpectedly.

Covenant design creates debt agreements that provide operational flexibility during shifting business scenarios while protecting lender interests. Traditional covenant structures can become constraining when business conditions change, requiring renegotiation or limiting operational options.

Rating agency considerations influence capital structure decisions based on credit-rating objectives and agency methodologies. Quantaland Utilities maintains investment-grade credit ratings through conservative capital structures that provide financing flexibility and lower financing costs.

International financing optimization considers currency exposure, regulatory requirements, and market access across multiple jurisdictions. Global companies can optimize financing costs and reduce risks through diverse financing sources and strategic currency hedging.

Market timing in financing decisions

Market timing enables companies to optimize financing costs and terms by issuing debt and equity when market conditions are favorable. While perfect timing is impossible, systematic market monitoring and flexible instrument selection can significantly improve financing outcomes.

Analysis of interest-rate cycles helps companies time debt issuance when rates are favorable relative to historical and expected future levels. Quantaland Manufacturing monitors interest-rate cycles and issues long-term debt when rates are below historical averages, while using short-term financing when rates are expected to decline.

Credit spread monitoring enables debt issuance timing based on corporate credit market conditions that affect financing costs independent of risk-free rates. Credit spreads vary significantly based on market sentiment and economic conditions, creating opportunities for astute issuers.

Equity market analysis guides equity issuance timing based on stock valuations, market sentiment, and investor preference for specific sectors. Companies can optimize equity issuance by timing offerings when valuations are attractive and investor appetite is strong.

Investor preference assessment helps companies time financing based on appetite for specific security types, industries, or credit qualities. Strong investor demand can support larger issuances at better terms, reducing overall capital costs.

Risk capacity versus risk appetite in leverage decisions

Leverage decisions require balancing their ability to service debt under stress with the willingness to accept financial risk for potential returns. This is essentially matching risk capacity with risk appetite. Quantum corporate finance maintains leverage flexibility that can adapt to changing risk conditions.

Stress-testing capabilities evaluate debt-service ability under various adverse scenarios, including revenue declines, margin compression, and increased costs. Quantaland Airlines conducts comprehensive stress testing that evaluates debt-service ability under scenarios including fuel price spikes, demand collapses, and regulatory changes.

Business model stability significantly affects leverage capacity. Businesses with stable cash flows can support higher leverage than those with volatile earnings patterns. Companies must assess their business model characteristics, including revenue predictability and margin stability, when determining appropriate leverage levels.

Industry comparison analysis examines leverage levels and financial performance across industry peers. This improves understanding of the appropriate leverage ranges for specific business types. These comparisons also provide context for leverage decisions while recognizing company-specific factors and competitive positioning.

Economic cycle considerations recognize that appropriate leverage levels may vary across economic cycles. Expansion periods may support higher leverage than recession periods, requiring companies to adjust their leverage based on economic cycle expectations and business outlook.

Financial flexibility preservation maintains debt capacity for future opportunities rather than maximizing current leverage. Quantaland Real Estate maintains conservative leverage that provides capacity for property acquisitions during market downturns when attractive opportunities emerge. At the same time, they are careful to protect against the fixed obligations that higher leverage would require.

Currency and interest-rate considerations in financing

International companies face complex financing decisions involving currency exposure, interest-rate environments, and regulatory requirements across multiple jurisdictions. These decisions require sophisticated approaches that consider multiple risk factors simultaneously.

Currency-matching strategies align the financing currency with the primary currency of cash flow to provide natural hedging against FX risk. Quantaland International finances its European operations with euro-denominated debt while financing its Asian operations with local-currency debt. This reduces their FX risk while optimizing local financing costs.

Analysis of the interest-rate environment considers interest-rate levels and volatility across different markets to optimize financing costs and risk. Companies can benefit from interest-rate differences between markets while managing interest-rate risk through strategic hedging approaches.

Regulatory arbitrage opportunities enable companies to optimize financing costs by accessing markets with favorable regulatory environments. Different jurisdictions may offer attractive financing terms to companies that meet specific criteria or operate in favored sectors.

Hedging strategy integration involves coordinating financing decisions with FX and interest-rate hedging strategies. Companies can use natural hedges through careful financing decisions and supplement them with derivative instruments when necessary to manage residual risks.

Tax optimization considers the impact of various tax treatments across multiple jurisdictions. Areas of consideration include interest expense deductibility and withholding tax implications. Quantaland Global optimizes its financing structure to minimize its overall tax burden while maintaining operational flexibility and regulatory compliance in all operating locations.

Covenant design for uncertain environments

Debt-covenant design balances lender protection with borrower flexibility, particularly during uncertain business conditions. Traditional covenant structures may become constraining when business conditions change, requiring renegotiation or limiting operational adaptability.

Financial ratio covenants must account for business volatility and potential changes in accounting standards or business mix. Quantaland Healthcare negotiates debt covenants with adequate cushions for seasonal cash-flow variations and potential changes in healthcare regulations that could affect financial performance.

Basket provisions provide flexibility for specific activities, including acquisitions, capital expenditures, and dividend payments, without requiring lender consent. These provisions enable businesses to adapt while maintaining overall debt protection for lenders.

Maintenance versus incurrence covenants create different restriction patterns. Maintenance covenants require ongoing compliance, whereas incurrence covenants apply only when specific actions are taken. The choice between these approaches significantly affects operational flexibility and financial planning.

Amendment procedures provide mechanisms for covenant modification when business conditions change significantly. Quantaland Energy includes covenant-flexibility provisions that account for commodity price volatility and potential changes in environmental regulations. This flexibility ensures that the company can adapt to market shifts without violating its debt agreements.

Alternative financing instruments and hybrid securities

Alternative financing instruments blend traditional debt and equity characteristics to optimize the cost of capital while maintaining preferred risk and return profiles. These instruments enable the financial structure optimization for specific business requirements and market conditions.

Convertible securities mix features of debt and equity, allowing for cheaper financing while letting investors benefit from potential gains. Quantaland Technology issues convertible bonds that have lower interest costs than regular debt but restrict conversion unless the stock performs well enough to make giving up some equity worthwhile.

Preferred stock provides equity capital with fixed dividend requirements that may be appropriate for companies needing permanent capital. This also provides investors with income assurance. Preferred stock can bridge debt and equity characteristics for specific financing needs.

Mezzanine financing provides capital between debt and equity that may be appropriate for growth companies requiring flexible capital. This option typically satisfies the demand for higher returns than debt but lower dilution than equity. Mezzanine capital often includes warrants or conversion features that provide additional upside potential for investors.

Asset-backed securities enable companies to monetize specific assets, including receivables, inventory, or equipment, through securities backed by these assets. These instruments can provide favorable financing costs while maintaining operational control over the underlying assets.

Contingent convertible bonds are loans that turn into shares of stock if certain events happen, such as when capital ratios drop too low. Quantaland Bank issues contingent convertible bonds that offer affordable capital and automatically change to equity if capital ratios fall below the required levels.

Dynamic distribution: adaptive dividends and buybacks

Dividend policies represent critical capital-allocation decisions that balance returning cash to shareholders with maintaining financial flexibility for business opportunities and contingencies. Quantum corporate finance approaches distribution policies as dynamic decisions that adapt to changing business conditions rather than following a rigid, predetermined formula.

Dynamic dividend policy frameworks

Traditional dividend policies often emphasize consistency and predictability, which can become inappropriate when business conditions change significantly. Dynamic dividend policies adapt distribution levels to evolving cash-flow patterns and business requirements while maintaining transparent communication with shareholders.

Cash flow–based policies link dividend payments to actual cash generation rather than accounting earnings, which may not reflect real cash availability. Quantaland Utilities employs cash flow–based dividend policies that adjust payments based on actual cash generation after capital expenditures. This approach maintains dividend sustainability while preserving flexibility during changing regulatory and business conditions.

Variable-dividend components combine base dividends with variable payments based on business performance and cash availability. These policies provide income predictability while enabling additional distributions during favorable business periods.

Dividend policy communication explains distribution frameworks to shareholders clearly, helping them understand the rationale behind the policies. Such communication also maintains management's flexibility to adapt to changing scenarios.

Special dividend capabilities provide mechanisms for distributing excess cash during strong business periods without committing to maintaining the same level in future distributions. Quantaland Manufacturing adjusts its dividend policies based on business cycle patterns that affect cash generation and planned capital investments. Their cycle-aware policies optimize shareholder returns while maintaining reinvestment flexibility for future growth opportunities

Share buyback timing and quantum market opportunities

Share buyback programs offer flexible mechanisms for returning cash to shareholders while providing management with discretion regarding timing and volume. Effective buyback programs can create significant shareholder value through strategic timing and execution approaches.

Valuation-based buybacks favor repurchase activity when stock prices appear undervalued relative to intrinsic value estimates. Quantaland Retail initiates share buybacks when their stock trades below estimated intrinsic value, avoiding buyback activity when valuations appear stretched. This value-oriented approach optimizes buyback effectiveness while preserving capital-allocation flexibility.

Market-timing strategies adapt buyback timing based on market volatility and liquidity conditions that affect execution success and the market. These strategies can improve buyback execution while reducing market disruption.

Opportunistic buyback programs keep buyback authorization in hand while making timing decisions based on market conditions and business requirements. These programs provide flexibility to capitalize on market opportunities while balancing other capital-allocation priorities.

Return policy frameworks balance share buybacks with dividend payments based on shareholder preferences and tax considerations. Quantaland Technology balances share buybacks with modest dividend payments to maintain total return flexibility. At the same time, they accommodate diverse shareholder preferences for income versus capital appreciation.

Cash-flow management through distribution policies

Distribution policies significantly affect corporate cash-flow management and financial flexibility. Effective policies must satisfy shareholder return expectations while meeting business financing requirements and preserving strategic optionality.

Cash-flow priority systems establish hierarchical priorities for cash usage, including debt service, capital expenditures, dividend payments, and share buybacks. Quantaland Mining prioritizes cash flow for debt service and maintenance capital expenditures before discretionary distributions, ensuring business sustainability while providing appropriate shareholder returns.

Seasonal-adjustment capabilities adapt distribution policies to seasonal business patterns that affect cash generation timing. These adjustments optimize distribution sustainability while accommodating business seasonality. A similar adjustment is necessary for cyclical businesses.

Policies for maintaining a liquidity reserve establish the minimum cash levels required for business operations and strategic flexibility before authorizing distributions. These policies ensure adequate liquidity while enabling distributions when excess cash is available.

Credit-facility coordination ensures distribution policies remain compatible with debt-covenant requirements and credit availability. This prevents distribution decisions that could trigger covenant violations or limit future borrowing capacity.

Investment-opportunity reserves maintain cash availability for strategic investments and acquisition opportunities while providing reasonable shareholder distributions. Quantaland Development maintains cash reserves for real estate acquisition opportunities while providing consistent, modest dividend payments, balancing current income with strategic investment flexibility.

Signaling effects and market communication

Distribution-policy changes send important signals to investors about management's confidence in business prospects and cash-flow sustainability. Understanding these signaling effects helps companies communicate effectively while managing market expectations.

Dividend initiation signals management confidence in business sustainability and cash-flow predictability. Quantaland Software declared its first dividend after achieving sustainable profitability and predictable cash flow, signaling a mature business model while maintaining investment capacity for growth.

Dividend increases signal management confidence in continued business growth and cash-flow improvement. Regular dividend increases can demonstrate business strength while creating expectations for ongoing growth.

Dividend reductions often signal business challenges or strategic changes requiring cash preservation. Companies can manage these signals through clear communication about business strategy and expected recovery timelines.

Communication on policy changes helps investors understand distribution modifications and their implications. Quantaland Services provides detailed explanations when modifying distribution policies, maintaining investor confidence through transparent communication.

Buyback announcements may signal management's belief that stock is undervalued or that the company has excess cash not needed for business operations. Large buyback programs often signal confidence in business prospects and stock valuation. This may trigger a valuation collapse unless handled carefully.

ESG considerations in capital return policies

Environmental, social, and governance factors increasingly influence distribution policies, as stakeholders expect companies to balance shareholder returns with broader stakeholder interests and sustainable business practices.

Stakeholder capitalism requires dividend decisions to balance shareholder returns with employee, customer, community, and environmental interests. Quantaland Metals balances dividend payments with investments in renewable energy, employee development, and community programs to optimize stakeholder value.

Environmental investment priorities may affect distribution policies. When companies allocate capital to sustainability initiatives, emissions reduction, or environmental remediation projects, these compete with dividend distributions for available cash. Similarly, social responsibility commitments, including employee benefits, community investment, and social-impact programs, may influence distribution policies. Companies need to prioritize social investments alongside shareholder returns.

Best practices in governance emphasize board oversight of distribution policies to ensure dividend decisions reflect an appropriate balance between shareholder returns and business-sustainability requirements. ESG-rating implications recognize that distribution policies affect ESG ratings, which in turn influence investor interest and financing costs. Companies may optimize distribution policies to maintain favorable ESG ratings while providing competitive shareholder returns.

Quantaland Mining adjusts its distribution policies to fund environmental remediation and sustainability initiatives while maintaining competitive returns. Their adjustment process demonstrates how quantum thinking enables adaptation to evolving stakeholder expectations.

Meanwhile, far from Quantaland

Zoom's adaptation post-pandemic (2022–2023)

The Quantum State: Zoom maintained financial optionality through the pandemic boom-bust cycle by building a significant cash reserve during peak demand. At the same time, they continued strategic investments as growth normalized, allowing simultaneous preparation for multiple future scenarios.

The Story: Zoom Video Communications' financial navigation of the post-pandemic transition represents a great example of quantum thinking in corporate finance. Zoom witnessed a dramatic pandemic-driven growth, with revenue jumping 326% year-over-year in Q2 2020 and its stock price increasing over 700% between January and October 2020. At the same time, Zoom anticipated a post-pandemic recalibration.

What distinguishes Zoom's approach is its ability to simultaneously harvest the financial benefits of pandemic-era growth while preparing for a more challenging future. During its peak growth phase in 2020–2021, Zoom maintained extraordinary capital discipline despite its sudden success. Rather than dramatically expanding its cost structure, the company allowed its operating margins to expand to over 40%, far exceeding software industry averages. This discipline resulted in Zoom accumulating a cash reserve exceeding $5.5 billion by January 2022, with zero debt. This created a financial fortress that provided quantum-like optionality as market conditions changed.

As pandemic restrictions eased and remote work partially normalized, Zoom's growth rate predictably decelerated from 355% in Q4 2020 to just 12% by Q3 2022. The stock price reflected this new reality, falling approximately 85% from its October 2020 peak to October 2022 lows. Yet despite this dramatic reversal, Zoom's financial position allowed it to pursue multiple strategic paths simultaneously. This was a classic quantum approach to uncertainty.

Zoom maintained a dual focus on maintaining profitability while investing in future growth. This reflected quantum thinking by simultaneously addressing near-term investor concerns and long-term competitive positioning.

Zoom's capital allocation during this transition period demonstrated this balance. The company initiated its first-ever share-repurchase program in February 2022, committing $1 billion to buying back shares at depressed valuations. This addressed investor concerns about the stock-price decline. Simultaneously, it increased R&D investments by 47% year-over-year in 2022 to develop new products beyond its core video-conferencing offering. These included contact-center solutions, AI-powered meeting assistants, and enhanced security features. The company also pursued strategic M&A. Zoom acquired the conversational-intelligence platform Solvvy in May 2022 while maintaining discipline on purchase prices. They reportedly took advantage of the tech downturn to negotiate favorable terms.

Perhaps most revealing was Zoom's employment strategy during this period. Many tech companies conducted massive layoffs in 2022–2023 after their aggressive hiring during the pandemic boom. In contrast, Zoom maintained a relatively stable headcount. They grew from approximately 6,800 employees in January 2022 to about 7,400 by January 2023, a strategic 8.8% increase. This measured approach reflected quantum thinking by neither dramatically expanding nor contracting. Instead, they maintained employee continuity while allowing natural attrition to manage costs.

By early 2023, this balanced strategy paid off. In Q4 2022, Zoom's growth remained modest at 4% year-over-year, while its operating margin stabilized at a healthy 34%. Their enterprise clients with yearly recurring revenue of $100,000 or more increased by 27% during this time. The company maintained its financial fortress with $5.4 billion in cash and marketable securities, providing continued optionality for both organic and inorganic growth opportunities.

The Quantum Finance Lesson: Zoom's approach to the post-pandemic transition serves as a strong example of quantum thinking applied to corporate finance. By keeping multiple strategic options open at once, Zoom was able to successfully navigate the unpredictability of its business environment. Just as quantum particles can exist in a superposition of states, Zoom maintained a balance between tightening costs and investing in growth, buying back its shares and pursuing acquisitions, and enhancing core offerings while exploring new markets.

This strategy reflects the quantum concept of tunneling, where particles overcome barriers that appear impassable in classical physics. Similarly, Zoom successfully navigated slowing growth by implementing multiple strategies simultaneously rather than restricting itself to a single course of action. Like a particle's probability cloud, Zoom's substantial cash reserves let the company "explore" various strategic options without being locked into any one path.

Through this multifaceted response to shifting market conditions, Zoom demonstrated a consistent strategic vision even when its tactical actions seemed contradictory.

Practical Application: Corporate finance leaders can incorporate quantum thinking by building and maintaining financial optionality that allows the pursuit of multiple strategic paths simultaneously. This is specifically relevant during periods of uncertainty. This entails maintaining stronger cash positions than are traditionally recommended while designing flexible cost structures that are ready for change.

Zoom preserved flexibility in their business model to meet the shifting market. This quantum perspective acknowledges that corporate success in uncertain environments requires maintaining probabilistic rather than deterministic strategies. They must prepare for multiple potential futures simultaneously.

Tesla's capital raising

The Quantum State: Tesla maintained a financial superposition between a high-growth technology company and a capital-intensive manufacturer. It leveraged its elevated stock valuation to raise funds when traditional debt financing would have been prohibitively costly.

The Story: Tesla's approach to capital raising exemplifies an innovative application of quantum thinking in corporate finance. Between 2017 and 2020, the electric-vehicle manufacturer successfully faced what many analysts considered a precarious financial position. They required massive capital investment to scale production at a time when they were generating consistently negative cash flow and carried significant debt. The company's solution demonstrated remarkable strategic flexibility by simultaneously appealing to growth and value frameworks rather than favoring one over the other.

The situation was particularly challenging in 2018–2019. Despite achieving breakthrough success with the Model 3, Tesla's production ramp created severe cash-flow pressures. In March 2018, Moody's downgraded Tesla's credit rating to B3, which was six levels below investment grade. The primary reasons cited were "significant shortfall in the production rate of the company's Model 3" and "liquidity pressures due to its large negative free cash flow." With approximately $1.2 billion in debt maturing by early 2019 and traditional financing options limited by its credit profile, Tesla faced a situation that was close to an existential threat.

CEO Elon Musk's public comments during this period reflected classic quantum thinking. He simultaneously acknowledged financial challenges while projecting technological confidence. In April 2018, Tesla sounded confident about being profitable with positive cash flow in Q3 and Q4. At the same time, they carried out management restructuring to improve communication and cut non-core activities, both vital to the success of their mission.

This dual messaging created a financial superposition that allowed Tesla to navigate between contradictory market narratives. To traditional lenders, Tesla emphasized its progress toward operational discipline and cash-flow positivity. To equity investors, it highlighted its revolutionary technology and long-term market opportunity. This approach proved successful when Tesla achieved its first quarterly profit in two years in Q3 2018, driving a 27% stock-price increase and creating new financing opportunities.

The culmination of this quantum strategy came in 2020 when Tesla conducted two opportunistic equity offerings. In February, with its stock having more than quadrupled over the previous six months, Tesla raised $2.3 billion through a public-stock offering. Remarkably, just six months later in September, Tesla raised an additional $5 billion through an at-the-market offering, with its stock having more than doubled again since February. These equity raises demonstrated Tesla's ability to leverage its elevated valuation to secure capital on terms that would have been impossible through conventional debt markets.

Perhaps most telling was that Tesla used this capital not for short-term operational needs but for strategic positioning. The funds accelerated factory construction in Berlin and Texas, increased R&D in battery technology, and strengthened Tesla's balance sheet by retiring high-interest debt. By 2021, this strategy had transformed Tesla's financial position. S&P upgraded the company's investment-grade credit rating to BB+. They also accumulated over $16 billion in cash reserves and established consistent profitability. Moreover, its strategic capital deployment during 2020–2021 positioned it to increase vehicle deliveries by over 87% in 2021, despite global supply-chain disruptions that severely constrained traditional automakers.

The Quantum Finance Lesson: Tesla's capital-raising strategy exemplifies how organizations can engage with contradictory market frameworks to maintain strategic flexibility. Much like quantum particles that exist in multiple states until observed, Tesla positioned itself both as a speculative growth company and an emerging industrial leader, adopting the identity that best suited the circumstances at any given time.

This approach, mirroring quantum theory, enabled Tesla to remain in financial superposition. They navigated the boundary between high-risk ventures and established manufacturers. By doing so, the company could tap into varied sources of capital, adapting to market conditions and investor perspectives. Tesla's unique market valuation and financial fundamentals operated in separate yet interconnected dimensions, allowing the company to harness market perceptions to enhance its core business realities.

Practical Application: Leaders in corporate finance should understand how to leverage market perceptions to enhance fundamental business positions. This means maintaining narrative flexibility that appeals to multiple investor frameworks while using favorable market conditions to create long-term strategic advantages.

As Tesla's experience demonstrates, quantum corporate finance isn't about manipulating perceptions. It is more about recognizing the interdependence between market narratives and business fundamentals. Companies that successfully navigate this interdependence can create positive feedback loops where improved perception enables improved reality. This in turn reinforces a positive perception.

Airbnb's shift during the pandemic

The Quantum State: Airbnb maintained strategic superposition during the pandemic by simultaneously contracting operationally while expanding its market vision. This allowed for rapid adaptation to radically changed travel patterns without abandoning its long-term mission.

The Story: Airbnb's response to the COVID-19 pandemic was one of the most dramatic examples of quantum-like corporate adaptation. Within weeks of declaring its intention to go public in 2020, the company faced an existential threat as global travel collapsed, with bookings declining by more than 70% and cancellations surging. This sudden reversal forced Airbnb to reconsider its entire business model while maintaining sufficient operational continuity to survive. The scale of the crisis was unprecedented. This shock required immediate action to preserve cash while the company's future business model remained deeply uncertain.

Airbnb's response demonstrated remarkable quantum-like thinking when they simultaneously pursued contraction and expansion strategies. On the operational side, the company implemented severe cuts. They laid off 25% of their workforce (1,900 employees), reduced their marketing budget by approximately 80%, and shelved numerous non-core initiatives. Executives also chipped in, and they accepted 50% pay cuts. Additionally, Airbnb raised $2 billion in emergency funding: $1 billion in equity from Silver Lake and Sixth Street Partners and $1 billion in debt on steep terms (10% interest plus warrants).

Yet even as Airbnb contracted operationally, it expanded conceptually! They reimagined their business model for a transformed travel landscape. Rather than treating the pandemic as a temporary interruption to normal operations, the company recognized it as a paradigm shift that would permanently alter travel patterns. They readied themselves for a new kind of travel market.

This reimagining took concrete form through rapid product innovation. In April 2020. Just weeks after implementing severe cuts, Airbnb launched Online Experiences, transforming its in-person hospitality offerings into virtual events hosted by guides worldwide. By June, the company was highlighting "nearby escapes" and rural destinations rather than urban centers. They had correctly anticipated the shift toward domestic travel and remote-work holidays.

Airbnb maintained coordination during this radical transformation remarkably well. Despite the massive operational changes, Airbnb emphasized cultural continuity by maintaining unprecedented transparency. They publicly shared the layoff memo explaining strategic decisions in detail. This document was widely praised for its clarity and empathy. Similarly, Airbnb provided regular public updates on marketplace trends, helping hosts adapt to changing traveler preferences.

This balanced approach of operational contraction and strategic expansion paid dividends when travel patterns eventually shifted in Airbnb's favor. By Q3 2020, rural bookings had grown from 40% to 60% of Airbnb's total business, and stays of 28+ days had become the fastest-growing category.

This development uniquely benefited Airbnb's model over traditional hotels. They proceeded with their IPO in December 2020 and achieved a first-day valuation of over $100 billion – more than Marriott, Hilton, and Hyatt combined.

The post-IPO period further validated Airbnb's quantum strategy. By Q2 2021, the company achieved its first profitable quarter since the pandemic began. Revenue also surpassed 2019 levels despite continued disruption in international and business travel.

The Quantum Finance Lesson: Airbnb's response to the pandemic shows how organizations can both contract and expand when faced with profound uncertainty. Much like quantum particles that occupy multiple states until observed, Airbnb preserved several strategic paths while quickly responding to shifting market realities.

This approach reflects the quantum principle of complementarity, in which seemingly opposing actions can both be valid ways of responding to a situation. By pursuing cost discipline while fostering innovation and balancing downsizing with forward-thinking expansion, Airbnb managed to sustain both survival and growth amid evolving circumstances. Their strategy was rooted in constantly reevaluating the business model, never assuming that pre-pandemic trends would return, and rapidly adapting to lasting changes in travel habits. In essence, Airbnb's methodology mirrored a quantum system that avoids "decoherence" by harmonizing immediate survival needs with the capacity for long-term transformation.

Practical Application: Business leaders facing existential disruption should distinguish between operational necessities and a strategic vision. This means implementing the required tactical contractions while maintaining or even expanding strategic possibilities. This complementary approach preserves both current viability and future optionality.

Airbnb prepared for a travel economy that was unrecognizable from pre-pandemic norms. This quantum perspective acknowledges that crisis periods collapse previous probability distributions. Such developments require leaders to harvest multiple potential futures simultaneously rather than awaiting a return to previous certainties.

Looking ahead

Quantum corporate finance transforms the CFO's role from optimizing static structures to orchestrating adaptive systems that thrive on uncertainty. By using real options for spending, changing financing methods, and adapting distribution plans, companies can keep their choices open while still providing steady value to stakeholders. Yet even the most sophisticated quantum corporate finance strategies face a fundamental challenge:

they must be implemented by human decision-makers whose psychology doesn't always align with rational frameworks. The intersection of human behavior and quantum market dynamics creates powerful forces that can either amplify or undermine the best corporate finance strategies. Next, we'll explore how psychological biases and cognitive patterns interact with quantum financial systems, revealing why understanding the human element is crucial for successful quantum finance implementation.

Chapter 12

Behavioral quantum finance

"*The glance of the tiger makes the deer run*," goes an old Quantaland proverb.

Day trader Zoe fell victim to the 2021 crypto hype. She had bought at the peak, driven by the fear of missing out (FOMO), only to lose half her capital when sentiment flipped. Her emotions, which dominated her decision-making, ultimately led to her downfall. This chapter describes how psychology creates quantum-like market swings. You will explore how observation influences behavior and delve deeply into the quantum mind hypothesis. The chapter will use cases such as crypto booms and the 2008 housing crisis to reveal entanglement effects. The lessons from the chapter will help you balance investor psychology and market dynamics.

Mind-market connection: psychology as a quantum force

Human psychology exhibits distinct quantum-like characteristics that interact with quantum financial markets in a complex and often unpredictable way. Understanding these psychological quantum effects is essential for explaining why financial markets frequently behave in ways that defy pure economic logic.

The intersection of quantum psychology and quantum finance creates powerful feedback loops where human consciousness actively influences market behavior, and market conditions simultaneously shape collective psychological states. These dynamic interactions generate emergent phenomena in market behaviors and psychological responses that go beyond what traditional psychology or market theory alone could predict.

The observer in the market: how analysis changes reality

Heisenberg's uncertainty principle reveals that observing quantum systems changes them in fundamental ways. Financial markets exhibit strikingly

DOI: 10.4324/9781003619635-12

similar characteristics. The very act of analysis and observation influences market behavior, creating an inherent uncertainty that cannot be eliminated through better information or more sophisticated analysis.

The observer effect in financial decision-making

Financial analysis creates quantum-like impacts where studying markets actively alters their behavior. Analysts' research and recommendations don't just describe market conditions. They actively reshape them through the power of observation.

Stock research demonstrates pure observer effects where analyst coverage influences trading behavior independent of fundamentals. When major investment firms initiate coverage of sparsely traded stocks, they boost trading volume and institutional interest regardless of whether the research sentiment is positive or negative.

Quantaland Research found their stock recommendations moved prices within hours of publication. Their analysis created buying pressure that eliminated the very undervaluation they had identified, forming feedback loops where the research became self-defeating through its own market impact.

Market-timing strategies show how widespread adoption alters the patterns that initially made the models effective. Technical indicators work until overuse invalidates their historical predictive power.

Economic forecasting creates impacts in which predictions shape outcomes. Recession forecasts can trigger defensive corporate behavior that actually increases the recession likelihood. Quantaland Research discovered that their quarterly forecasts influenced investment decisions, ultimately shaping the economic conditions they were trying to predict.

Central bank communication triggers institutional impacts where policy statements influence markets before any actual policy changes occur. This shows that observation and anticipation can be as powerful as concrete action.

Measurement uncertainty in market analysis

Financial measurement faces quantum-like imprecision, where precisely measuring certain market traits reduces accuracy in measuring others. This creates fundamental trade-offs that no analytical tools can eliminate.

Price discovery shows this imprecision clearly. Efforts to pinpoint security values inevitably influence both trading behavior and the values themselves. The measurement process alters what's being measured.

High-frequency trading may improve short-term price precision but often reduces liquidity and increases volatility. Quantaland Trading found that their pricing algorithms worked well in normal conditions but caused instability during stress periods. Their precise analysis tools altered market microstructure in ways that eventually reduced their intended accuracy.

Risk analysis exhibits similar imprecision, where precise historical risk measures reduce accuracy in predicting future patterns. VaR models based on historical data often fail catastrophically during crises when correlation patterns shift dramatically.

Performance attribution faces measurement uncertainty, where identifying sources of past returns may affect future performance. Managers who understand what worked historically may overuse those strategies, reducing their future effectiveness.

Quantaland Risk Management discovered that short-term volatility models worked for daily needs but failed for strategic planning. They now use different models for varying time horizons instead of seeking a single analytical solution.

Confirmation bias and selective observation

Confirmation bias creates quantum-like effects where analysts preferentially notice evidence supporting their initial hypotheses while ignoring contradictory data. This systematic distortion shapes both research design and information gathering for investment decisions.

Research design practices influence how analysts structure their investigations to favor confirming evidence. They tend to ask questions and gather data that support initial impressions rather than pursuing truly objective analysis.

Quantaland Analysis found that their analysts gave disproportionate weight to sources that supported their stock recommendations. This behavior skewed purported objective analysis toward predetermined conclusions.

Information source selection also shows analysts' tendency to prefer sources supporting their views while avoiding contradictory ones. This ends up creating echo chambers that reinforce existing beliefs.

Data interpretation tendencies affect how analysts process ambiguous information based on existing beliefs. Ambiguous earnings and market data can often support almost any conclusion, making objective interpretation particularly challenging.

Team members were required to argue against investment decisions as part of Quantaland Investments' devil's advocate procedures. This procedure greatly lessened confirmation bias in their decision-making by requiring them to weigh conflicting information.

Overconfidence in prediction accuracy

Confidence bias creates quantum-like effects where investors believe they can predict market outcomes more accurately than what the evidence supports. This overconfidence influences behavior in ways that often reduce returns while increasing risk exposure.

Precision illusion emerges when sophisticated analytical tools create false confidence in prediction accuracy. Complex models provide the impression of precision that exceeds their actual capability, leading to overly confident decisions.

Quantaland Forecasting found that detailed models encouraged managers to take larger positions and trade more frequently on the basis of precise-seeming analysis. The tool's sophistication increased confidence without actually improving prediction accuracy.

Hindsight bias reinforces overconfidence by making past events seem more predictable than they actually were. Successful decisions appear obvious in retrospect, creating false confidence in predicting future outcomes.

Attribution bias shapes how investors interpret results to maintain confidence. Success is typically credited to skill, while failures are blamed on adverse luck, leading to the systematic overestimation of predictive ability.

Quantaland Research implemented calibration exercises where analysts assessed their prediction confidence and then compared it to actual outcomes. This self-analysis reduced unfounded confidence and significantly improved decision quality.

The illusion of knowledge and control

Financial professionals often develop beliefs that information and analysis can control uncertain outcomes. These beliefs create quantum-like impacts as attempts to increase control may actually lead to greater uncertainty and increased risk exposure.

When we believe that large amounts of data improve prediction capability, information overload creates false assurance. More information often increases confidence without actually improving accuracy.

Quantaland Data Services found that managers with comprehensive data showed more confidence in their decisions but didn't actually outperform those working with simplified summaries. Information quantity boosted assurance without leading to better decisions.

Analysis sophistication effects emerge when complex techniques create impressions of precision in inherently uncertain domains. Sophisticated models may explain historical relationships beautifully, without necessarily improving future predictions.

Control attribution tendency leads investors to believe that they have more mastery over outcomes than the markets actually allow. This encourages active strategies and frequent adjustments that often reduce returns while increasing costs.

Quantaland Technology discovered that sophisticated algorithms created false confidence, leading traders to increase position sizes based on illusory beliefs in their expertise. They now set position limits based on actual performance rather than perceived analytical sophistication.

Cognitive dissonance in uncertain markets

Cognitive conflict occurs when investors hold contradictory beliefs or when they clash with market evidence. Rather than resolving these through rational analysis, investors often maintain contradictions in a quantum-like coexistence until events force resolution.

Market efficiency contradictions show investors believing markets are efficient while simultaneously trying to achieve superior returns. Quantaland Advisers found that the very clients who expressed belief in market efficiency were the ones who demanded active strategies to beat the market. They held both beliefs until results forced them to confront the inconsistency.

Risk tolerance inconsistencies create conflict where investors claim high-risk tolerance in good conditions but show strong aversion during market stress. These preference contradictions persist until market conditions force actual risk decisions.

The diversification paradox shows investors understanding diversification benefits intellectually, while emotionally favoring investments in familiar assets. Knowledge and behavior contradict until market events force alignment.

Conflict in performance attribution emerges when results contradict expectations. Rather than updating beliefs, investors often maintain contradictory explanations until overwhelming evidence forces a change.

Quantaland Behavioral Studies found investors could hold conflicting market beliefs for extended periods, typically resolving these inconsistencies only through emotional or financial crises rather than through logical analysis.

The quantum brain: decision-making beyond classical logic

The quantum mind hypothesis proposes that human consciousness and decision-making operate according to quantum principles rather than classical logic alone. While controversial, this perspective helps explain aspects of financial behavior that traditional psychological models struggle to address comprehensively.

Quantum processes in human consciousness and decision-making

Human decision-making can be modeled using quantum-like characteristics, such as superposition (considering multiple possibilities simultaneously), entanglement (correlated mental states), and nonlocality (influences beyond direct communication). While these are allegories rather than descriptions of literal quantum processes, financial decisions exhibit patterns that can be effectively modeled using quantum mathematical frameworks. If we better understand subconscious behavior, it could shape investment decisions beyond classical psychological explanations.

Superposition in decision-making allows humans to hold multiple potential choices simultaneously until external events collapse them into a specific decision. Investment choices often exist across multiple possibilities until market conditions force a particular outcome. Researchers from Quantaland Psychology believe that investors frequently maintain contradictory intentions, such as optimism and caution about a stock. They hold such views until external events prompt a final decision, making a purely linear, logical progression less central to the outcome. This aligns with quantum cognition models, though direct evidence in consciousness remains speculative.

Quantum-like coherence in consciousness may enhance decision quality when mental states achieve coherent alignment. Meditation-induced mindsets and flow states appear to improve financial decision-making through enhanced pattern recognition and intuition, as supported by studies on mindfulness in cognitive performance. Further research would provide a deeper understanding of this psychological process that mirrors quantum behavior.

The concept of connection effects in consciousness may suggest nonlocal correlations among decision-makers, potentially influencing market behavior in ways that seem to transcend classical communication channels. It may be interesting to study whether traders exhibit distinctive brainwave patterns during decision-making. It is critical to emphasize, however, that these patterns most likely reflect conventional neural synchronization, which could still explain the apparent quantum-like behavior.

Superposition of mental states in financial decisions

Investment decisions frequently involve quantum-like coexistence, where multiple potential choices exist in consciousness until events collapse the possibilities into specific actions. This explains investors' genuine experience of uncertainty until a commitment occurs.

Ambivalent investment states demonstrate this coexistence, where investors simultaneously want to both buy and sell a security until market conditions force an action. Quantaland Decision Research found investors often experience measurable stress when holding positions that they want to both keep and sell. This suggests that these desires represent genuine coexistence states rather than simple logical conflicts leading to indecision.

Multiple-scenario consideration involves maintaining several investment approaches simultaneously until market conditions determine the optimal one. Successful investors typically maintain this superposition longer than those who collapse their options too prematurely.

Intuitive decision integration combines analytical and emotional inputs in coexistence until decision moments force specific choices. Decisions often benefit from holding both perspectives rather than choosing one exclusively. Option paralysis reflects excessive coexistence, where too many possibilities exist without effective collapse mechanisms, leading to analysis paralysis and decision avoidance.

Decision-timing sensitivity shows how external events trigger coexistence collapse, often yielding varied outcomes from similar initial states. Quantaland Investment discovered that making identical decisions at different times resulted in significantly different outcomes, which suggested nonclassical timing sensitivity instead of logical optimization.

Entanglement effects in collective investor psychology

Collective investor behavior may exhibit quantum entanglement, where individual decisions correlate through nonclassical mechanisms that extend beyond traditional information sharing and social influence.

Crowd synchronization in market behavior may reflect quantum entanglement. In these instances, the mental states of investors correlate in ways that cannot be explained by classical mechanisms alone. Market bubbles and crashes often involve entanglement effects that create collective behavior exceeding what individual decisions would predict.

Quantaland Behavioral Society research discovered decision-making patterns that exceeded classical information-sharing models, suggesting possible quantum entanglement in collective psychology. Sentiment contagion beyond classical communication may involve quantum entanglement, where investor emotions correlate without direct interaction. This explains how market sentiment propagates through seemingly nonclassical channels.

Geographic correlation anomalies may indicate quantum-like entanglement, where investors in different locations show correlated patterns that can be modeled mathematically using quantum-inspired frameworks.

Quantum coherence and decoherence in market sentiment

Market sentiment may exhibit quantum coherence during periods of strong consensus and decoherence during uncertainty phases. Understanding these patterns reveals the dynamics of market stability and instability.

Coherent market phases show collective investor actions creating stable trends and predictable behavior. During these periods, decisions align to produce strong directional movements with low volatility.

Quantaland University identified periods with unusual consensus and predictable price movements that might reflect quantum coherence in market psychology. Their research suggested distinctive coherence patterns in sentiment data.

Decoherent market phases display collective disagreement and uncertainty, creating volatile, unpredictable behavior. Decoherence typically precedes major market transitions when consensus breaks down.

Coherence breakdown events trigger rapid shifts from stable to unstable conditions when market agreement dissolves. Quantaland Media Studies found that financial news influenced sentiment coherence, with media consensus creating effects that extended beyond information content alone.

Nonlocality in financial information processing

Quantum nonlocality in financial information processing can influence decisions without a direct causal connection. These effects help explain market anomalies that defy classical models.

Instantaneous market correlation across geographically separated regions may resemble quantum nonlocality. Although high-frequency trading and electronic communication systems can explain the price movements, the situation resembles a quantum-like environment.

Quantaland Global Markets' research revealed correlation patterns between distinct markets that communication speed alone could not explain. This raises a question of whether these nonlocal correlation patterns can be modeled using quantum-inspired mathematics.

Anticipatory market movements before information releases may involve quantum nonlocality, where markets react before data become publicly available. Conventional insider trading models cannot explain all anticipatory behavior patterns.

Cross-asset correlation anomalies may indicate nonlocal correlations where price movements correlate without fundamental economic ties. Quantaland Analytics found patterns suggesting market behavior correlates with future events in ways that exceed classical models, indicating nonlocal patterns in information processing.

Quantum tunneling in breakthrough investment insights

Investment breakthroughs and creative problem-solving may involve quantum tunneling-like processes. This is where understanding emerges suddenly, bypassing classical analytical steps.

Eureka moments in investment analysis may reflect nonclassical leaps, where complex understanding emerges without logical progression. Successful investors frequently report insights appearing suddenly rather than through systematic analysis, suggesting nonclassical effects in financial intuition.

Pattern-recognition breakthroughs may involve nonclassical processing, where complex market patterns become clear beyond standard analytical capabilities. Market timing and trend identification often benefit from these nonclassical recognition processes.

Contrarian insights frequently involve nonclassical leaps, where investors identify opportunities that defy conventional wisdom. These insights typically emerge through cognitive processing that transcends logical analysis alone.

Quantaland Strategic Consultants found that successful investment managers often developed their business philosophies through sudden insights that deviated from classical rational models. The use of nonlogical cognitive processes in exceptional investment decision-making can explain this.

Biased measurements: mind over money

Cognitive biases interact with quantum market conditions to create complex decision-making challenges that traditional bias-mitigation techniques cannot fully address. Understanding these interactions helps develop more effective approaches to bias management in dynamic financial environments, where reference points constantly shift.

Anchoring effect

Anchoring bias creates quantum-like reference points that influence all subsequent financial decisions. These anchors extend beyond individual choices to create persistent reference frameworks that shape entire investment strategies and risk-management approaches. These approaches often persist even long after they've become irrelevant.

Price anchoring in investment valuations

Purchase prices create powerful reference points that influence investment decisions long after the initial transaction. These benchmarks shape holding periods and performance evaluations, often contradicting logical analysis and changing market conditions.

Loss aversion bias occurs when investors refuse to sell below the purchase price, ignoring changing fundamentals. The original cost creates a reference point that dominates future decisions, despite maintaining suboptimal positions. Quantaland Advisers found that clients refused to sell stocks below purchase prices, even when better investment opportunities were available, showing how this bias overrode logical portfolio optimization.

Decision updating becomes difficult when investors cannot adjust internal benchmarks to reflect changed market conditions. Often, investors judge stocks from bubble periods against past high prices, years later, which hinders rational decision-making.

Gain realization bias causes investors to be eager to sell positions that have increased in value compared to their purchase price, while they tend to hold onto investments that are losing value. This systematically tilts portfolios toward underperforming assets.

Quantaland Taxation Services discovered that clients exhibited this bias even in tax-advantaged accounts, where tax consequences were irrelevant. This mindset demonstrates how anchoring transcends rational optimization.

Historical performance anchoring in expectations

Past performance creates reference points that influence return expectations and risk tolerance beyond what current market conditions justify. These benchmarks persist despite fundamentally changed environments.

Bias of peak performance occurs when investors base expectations on historical peaks rather than on average or prospective performance. Technology investors in the late 1990s expected similar returns years after the markets had fundamentally changed.

Recession bias creates effects where investors expect poor performance after downturns, persisting despite improved economic conditions. This often leads to excessive equity avoidance due to past losses.

Generation-specific bias shows that different age cohorts develop varied expectations from historical experiences. Baby Boomers with high inflation experience differ significantly from younger investors with different market exposures.

Quantaland Recruitment found that investors were stuck with underperforming managers who had strong historical records. These investors were allowing past performance to override current evidence.

Reference point bias in portfolio management

Portfolio management decisions become tied to reference points such as benchmarks, peer performance, and historical values. These benchmarks shape risk tolerance and allocation decisions, often misaligning with actual investment goals.

Benchmark bias creates portfolio decisions aimed at minimizing tracking errors rather than optimizing absolute returns. Quantaland Portfolio found that managers avoided positions that would create benchmark deviation, even when analysis suggested better risk-adjusted returns were available. This bias dominated optimal portfolio construction.

High-water mark bias affects manager behavior when performance is measured against past peak portfolio values. This creates perverse incentives for either risky recovery strategies or overly conservative gain protection.

Peer performance bias influences decisions based on relative manager performance rather than absolute returns. This leads to herding behavior and avoidance of strategies that might risk short-term underperformance. Quantaland Funds found that managers maintained cash levels near historical averages despite market factors suggesting different allocations would better serve investment goals.

Adjustment insufficiency from initial estimates

Investment analysis often suffers from bias where initial understanding creates reference points that subsequent analysis fails to adequately adjust. This tendency favors initial impressions despite new information suggesting different conclusions.

First impression bias in security analysis occurs when initial research and price targets create benchmarks that subsequent analysis cannot overcome. Quantaland Research found that price targets set by analysts showed inadequate adjustment from initial estimates, despite substantial new information meriting significant changes. Initial estimates created anchors that analysis couldn't fully overcome.

Starting point bias shows how initial assumptions in financial models create results leaning toward those points. Discounted cash-flow models particularly reflect bias toward initial growth and discount rate assumptions.

Budget bias also affects corporate planning, where initial proposals create reference points for final allocations. Management often tweaks numbers to ensure results match initial assumptions rather than reflecting current reality.

Quantaland Industry Analysis found that companies' decisions clustered around industry averages, even when company-specific analysis suggested different approaches. A strong bias toward industry benchmarks was very evident.

Cultural and social anchoring influences

Cultural background and social environment create powerful biases that influence financial decisions through family, community, and cultural reference points. These benchmarks operate below conscious awareness, significantly shaping investment behavior.

Family financial experience creates bias where investment decisions reflect family history and attitudes toward money, risk, and investment. These benchmarks persist across generations despite differing individual circumstances.

Quantaland Cultural Studies found that risk tolerance and allocation preferences correlated strongly with family and cultural heritage. These patterns held even among individuals with similar income and education levels, showing powerful bias based on social reference points.

Geographic bias influences investment decisions based on local market experience and regional conditions. Investors consistently show home country bias, favoring familiar geographic reference points over global diversification opportunities.

Social class bias affects behavior through peer group norms and status considerations. Investment strategies often reflect social comparisons rather than financial optimization.

Quantaland Social Psychology found that investment decisions correlated strongly with media and social media exposure. This underlines how external communication creates bias, influencing portfolio decisions.

Temporal anchoring in financial planning

Financial planning becomes tied to specific times and life events, creating reference points for saving, spending, and investment decisions. These benchmarks may not align with changing life circumstances and market conditions.

Retirement age bias occurs when planning fixates on specific ages rather than on optimal timing based on financial readiness. Traditional retirement ages create arbitrary benchmarks that don't reflect individual circumstances.

Quantaland Advisers found that clients planned exclusively for age-65 retirement, even when analysis suggested earlier or later retirement would better suit their circumstances and preferences. This fixation disregarded life events that required greater attention than mere retirement timing.

Education funding bias creates reference points around traditional college costs and fixed-year models. In reality, this may not align with optimal education investment strategies in an era of diverse educational alternatives.

Home purchase bias influences housing decisions based on traditional timing and price expectations, which may not optimize financing decisions and investment strategies. Social expectations create additional bias affecting housing investment decisions.

Quantaland Advisers found that clients chose allocations matching traditional age-based norms, while individual risk tolerance and circumstances suggested different approaches would be more appropriate.

Herd behavior as a quantum phenomenon

Herd behavior in financial markets creates quantum-like entanglement effects, where individually rational decisions combine to produce collectively irrational outcomes. Understanding these herding mechanisms helps explain market bubbles, crashes, and other phenomena that analysis of individual decisions cannot predict.

Information cascades and social proof in markets

Information cascades occur when investors ignore private information and follow others' decisions. This creates quantum-like behavior, where early choices disproportionately influence subsequent actions. Sequential decision-making creates cascades where early investors' choices signal information that influences others, even when the initial data are poor or incomplete.

Quantaland Investments studied trading patterns and found that early-morning decisions influenced trading throughout the day, even when afternoon data contradicted the morning signals. Early decisions created cascades that persisted despite contradictory information.

Social proof mechanisms drive herding when investors interpret others' actions as signals of an optimal strategy. Observing others buying particular securities, they infer superior information and follow suit, often overlooking evidence to the contrary.

Expert opinion cascades occur when respected analysts' decisions spark widespread imitation despite opaque analytical foundations. Their opinions create cascades that persist even when subsequent analysis provides better information.

Quantaland Media found that investment themes with heavy media coverage drew excessive capital flows that exceeded what fundamental analysis would justify. It was evident that cascade effects driven by media attention were more powerful than the fundamental merit of the investment.

FOMO and investment bubbles

FOMO creates psychological pressure that drives bubble formation through quantum-like feedback cycles. Rising prices attract more buyers, despite fundamentals not supporting such valuations, creating self-reinforcing momentum. Momentum cycles emerge when price increases attract buyers fearing missed opportunities. These cycles persist beyond reasonable valuations, as psychological momentum overrides fundamental analysis.

Quantaland Analytics documented FOMO effects during technology bubbles, where investors bought at increasingly higher prices, fearing missed gains while overlooking deteriorating fundamentals. Social comparison also drives FOMO when investors see others reporting superior returns and feel pressure to match strategies. This creates overvaluation, as social comparison overshadows analytical judgment.

Regret avoidance strongly motivates FOMO, where investors dread missing opportunities more than they fear losses from overvalued investments. This psychological bias drives overinvestment in emerging bubbles. Lottery-like investments spark FOMO that supports demand for high-return, low-probability assets. Technology stocks and cryptocurrencies often attract disproportionate investment based on FOMO rather than expected value analysis.

Professional FOMO affects institutional investors who feel pressure to join popular themes to avoid career risks. By channeling significant institutional capital flows into well-liked themes, this phenomenon intensifies market bubbles.

Quantaland Professionals found that managers increased allocations to popular themes even when their analysis showed overvaluation. Fear of career underperformance, rather than investment merit, drove decisions.

Contrarian investing and crowd psychology

Contrarian investment strategies exploit herd behavior by taking positions against prevailing crowd sentiment. Understanding crowd psychology helps identify mispricing opportunities that these strategies can exploit.

Extreme sentiment identification focuses on spotting periods when crowd sentiment reaches peak levels, often signaling impending market reversals. Contrarian strategies work best when consensus becomes overwhelming and alternative views are scarce. Quantaland Strategies developed systems to identify extreme consensus periods. By identifying these signals, they were able to achieve risk-adjusted returns that surpassed those of following crowd sentiment.

Crowded trade recognition identifies when popular strategies attract excessive capital, reducing future return potential. Overcrowded trades typically underperform, as too many investors pursue similar approaches simultaneously.

Market timing via crowd psychology uses sentiment indicators for entry and exit decisions based on behavioral patterns. Extreme optimism often signals market peaks, while extreme pessimism suggests a potential bottom.

Quantaland Value Investing found that successful contrarian strategies required long holding periods, as crowd sentiment often persisted longer than predicted. This necessitated portfolio structures that could sustain positions without premature liquidation.

Social media amplification of herd behavior

Social media platforms dramatically amplify herd effects through rapid information sharing and viral content distribution. Algorithmic recommendations accelerate crowd formation and intensify collective behavior. Viral themes spread on social media faster than traditional financial media can evaluate them, creating rapid crowd development around investment ideas. These drive capital flows ahead of proper merit analysis. Quantaland Media found that themes gaining viral traction attracted significant capital flows within days, while professional analysis lagged in assessing their actual merit or risks.

Algorithmic amplification occurs when social media algorithms promote high-engagement content, favoring sensational themes over balanced analysis. This accelerates herd behavior by amplifying extreme viewpoints.

Echo chamber effects create environments where users primarily see content that reinforces their existing investment views. These chambers intensify conviction and filter out contradictory data that might moderate herd behavior.

Influencer impact demonstrates how personalities drive herd behavior through recommendations to large follower bases. Social media influencers can drive market movements primarily through their social reach instead of their investment expertise.

Institutional herding and benchmark hugging

Professional investment managers often exhibit stronger herd behavior than individual investors. Career risks and institutional constraints often dictate this, favoring conventional decisions over optimal strategies.

Career risk management drives institutional herding as managers fear underperforming compared to their peers more than achieving suboptimal returns. This creates strong incentives to adopt conventional strategies that match competitors' approaches. When researchers with Quantaland Research prefer strategies that minimize career risk rather than maximize returns, that leads to systematic herding among professionals

Benchmark alignment occurs when managers focus on minimizing errors rather than optimizing absolute returns. This creates herding around benchmark holdings rather than pursuing optimal opportunities. Relative performance evaluation creates incentives favoring peer comparison over absolute returns, resulting in encouraging herd behavior. These systems create structural bias against contrarian strategies that might underperform temporarily.

Quantaland Consultant found that institutions using similar consultants showed correlated manager and strategy choices, demonstrating herding effects arising from shared advisory relationships rather than independent analysis.

Network effects on investment decision-making

Investment networks create information-sharing and social-influence patterns that drive herd behavior. Professional relationships and affiliations often influence investment decisions more than objective analysis.

Professional network influence occurs when investment professionals share information through conferences, associations, and informal relationships. These interactions create a consensus that influences decisions across multiple institutions. Quantaland Research mapped professional relationships among investors and found decision correlations reflecting network ties rather than independent analysis.

Geographic clustering creates regional herd patterns, where professionals develop similar approaches through local networking and information sharing. Different financial centers often exhibit distinct investment styles emerging from network effects.

Then you have educational-background influence. You are likely to find professionals from similar institutions sharing similar investment philosophies. Alumni networks also create herd effects that persist throughout careers.

Conference circuit influence occurs when professionals hear similar presentations at industry events, forming consensus around particular investment themes. Conference networking drives herd behavior as ideas spread through personal interactions.

Quantaland analysts found that themes emphasized at major conferences consistently drew increased attention and capital flows. These are clear signs of herd effects arising from conference attendance rather than independent analysis.

Meanwhile, far from Quantaland

FOMO during the cryptocurrency boom and bust (2020–2022)

The Quantum State: Cryptocurrency markets exhibited quantum-like entanglement between price movements and investor psychology. The FOMO created self-reinforcing feedback loops that accelerated both boom-and-bust cycles.

The Story: The cryptocurrency market's extraordinary boom-bust cycle between 2020 and 2022 provides a compelling case study in behavioral quantum finance. It showed how psychological factors can create seemingly irrational yet self-reinforcing market dynamics. During this period, Bitcoin's price surged from approximately $5,000 in March 2020 to nearly $69,000 in November 2021 before crashing below $16,000 by November 2022. At the same time, the broader cryptocurrency market experienced even more dramatic volatility.

What made this cycle particularly instructive was the explicit role of FOMO in driving market behavior. Traditional markets often mask psychological drivers with rationalized narratives. Interestingly, many cryptocurrency participants openly acknowledged that the basis of their investment decisions was psychological and not analytical. Many investors who missed the Bitcoin run thought that they would see the same run with Dogecoin.

This FOMO-driven behavior created quantum-like feedback loops between market prices and investor psychology. Rising prices triggered media coverage, which increased public awareness, which drove new investor participation, which further increased prices. This self-reinforcing cycle accelerated between 2020 and 2021. The FOMO was influencing a large number of investors.

The phenomenon extended beyond Bitcoin to newer cryptocurrencies with low fundamental value propositions. Dogecoin, created as a joke in 2013, saw its market capitalization exceed $88 billion in May 2021 after social media promotion by figures like Elon Musk. Shiba Inu, a "meme token" launched in 2020 explicitly as a Dogecoin derivative, achieved a market capitalization of over $40 billion by October 2021. You may start thinking that the barter system had a better basis for valuation than the crypto.

Perhaps most revealing was how FOMO infected institutional investors traditionally assumed to be more rational market participants. Survey data from Fidelity Digital Assets found that 70% of institutional investors planned to buy or invest in cryptocurrencies within five years, despite many lacking clear justification beyond price momentum. Even sophisticated financial institutions like Goldman Sachs and Morgan Stanley began offering cryptocurrency exposure to wealth management clients in 2021. Their timing coincided with market peaks.

The role of social media in amplifying FOMO created particularly quantum-like market dynamics, where observation directly influenced the system being watched. The more social media discussed Bitcoin, the more people took an interest, suggesting a reflexive relationship between social attention and market performance.

This FOMO dynamic reversed dramatically during the 2022 downturn. Falling prices generated negative publicity, leading to reduced investor interest, which in turn triggered further selling and accelerated price declines. Volatility continues to hinder institutional adoption. The wild boom-and-bust cycles hinder institutional adoption.

The Quantum Finance Lesson: The impact of FOMO on cryptocurrency markets demonstrates the interplay between investor psychology and asset prices, a phenomenon known as behavioral quantum effects. These feedback loops drive boom-and-bust cycles, as observation of price movements directly impacts investor choices and, in turn, shapes those price movements.

This dynamic closely reflects the quantum principle of the observer effect, where the act of observing a system changes its actual state. In the world of cryptocurrencies, monitoring prices was not passive; rather, it actively influenced outcomes through psychological feedback. Investors became participants in the creation of price movements, with trading decisions driven by psychological factors rather than solely by objective analysis.

Practical Application: Investors should understand how psychological factors create feedback loops that can accelerate market movements. This means monitoring sentiment indicators along with fundamental metrics and being aware of how collective psychology influences individual decision-making.

This quantum perspective acknowledges that market narratives, like quantum states, can maintain coherence despite limited evidentiary support. It creates both risks and opportunities for investors who recognize these dynamics.

Dot-com bubble (late 1990s)

The Quantum State: Internet companies existed in a superposition of speculative valuation and fundamental business reality. Market psychology collapsed this duality when investor sentiment shifted from optimism to skepticism about untested business models.

The Story: The dot-com bubble of the late 1990s showed how market psychology can create quantum-like states of valuation that temporarily defy fundamental business realities. Between 1995 and 2000, the NASDAQ Composite Index rose from under 1,000 to 5,048 on March 10, 2000. Internet-related stocks, with limited revenues and often no profits, primarily drove this 400% increase. This extraordinary valuation expansion, followed by an equally dramatic collapse, illustrates how behavioral factors can create and then destroy market value. Classical financial valuation theory never expected such dramatic movements.

The bubble's formation exhibited classic quantum-like characteristics. Uncertainty played a major role in allowing multiple contradictory valuations to coexist. As the internet emerged as a transformative technology in the mid-1990s, genuine uncertainty about its commercial potential created space for wide-ranging growth projections. This uncertainty allowed companies with minimal financial histories to achieve astronomical valuations based almost entirely on narrative rather than on fundamentals.

Very few companies exemplify this phenomenon better than Pets.com. Founded in 1998, the online pet-supply retailer became a public company in February 2000 through an IPO that raised $82.5 million. This occurred while the company had generated only $5.8 million in revenue and incurred $61.8 million in losses the previous year. The company's famous sock-puppet mascot and Super Bowl advertising created brand recognition that temporarily substituted for business viability. This space allowed its market capitalization to reach hundreds of millions despite a fundamentally flawed business model. Pets.com would cease operations just nine months after its IPO, liquidating its assets and becoming an iconic symbol of the bubble's excesses.

Similar patterns appeared across numerous internet companies. Webvan, an online grocery delivery service, raised over $800 million in venture capital and public financing. They collapsed in 2001 after burning through cash at an unsustainable rate. Kozmo.com promised one-hour delivery of small items without charging delivery fees, raising over $250 million before shutting down. The internet enabled unprecedented speculation that allowed the creation of wealth at an unimaginable pace.

What made these valuations possible was a collective suspension of acting on traditional financial metrics. Alternative measurements such as "eyeballs," "mindshare," and "first-mover advantage" replaced historical valuation methods based on earnings, cash flow, or even revenue. The irrational exuberance persisted for a long period as new metrics temporarily displaced traditional valuation frameworks.

The psychological feedback loops that sustained the bubble reflected quantum-like entanglement between investor behavior and market prices. Rising valuations attracted media attention, which drove retail investor participation, which further increased prices. This created a self-reinforcing cycle disconnected from business fundamentals. Market prices are known to reflect expectations, and those same expectations can end up influencing the so-called basis of valuation.

The bubble collapse began in March 2000 and accelerated through 2001–2002, with the NASDAQ ultimately losing approximately 78% of its value. This dramatic reversal demonstrated how quickly market psychology could shift from euphoria to panic. In the process, it collapsed superpositions of value into more definitive states based on fundamental business metrics.

The dot-com bubble's aftermath revealed the distinction between genuine innovation and speculative excess. While many dot-com companies failed entirely, others with viable business models eventually recovered and prospered. Amazon.com is one of the apt examples. They saw their stock price fall from $107 in December 1999 to $5.97 on September 28, 2001, but ultimately became one of the world's most valuable companies. This divergence illustrated how the bubble contained both legitimate transformational potential and unsustainable speculation. This resembled a quantum-like superposition that market forces eventually resolved through a painful but clarifying correction.

The Quantum Finance Lesson: The dot-com bubble is a clear example of how financial markets can exist in quantum-like superpositions of value during times of genuine technological uncertainty. Just as quantum particles remain in multiple possible states until measured, emerging tech firms can temporarily support a range of valuations. Eventually, the market evaluates these companies based on changes in psychology or investor sentiment. This forces values to align with business realities.

During the dot-com era, company valuations exhibited a wave-like distribution of possibilities. They lasted until changes in market psychology collapsed those multiple outcomes into concrete assessments tied to actual business fundamentals. This mirrors quantum mechanics' wave-function collapse, where an array of probabilities becomes a single, definite reality when observed.

Ultimately, dot-com valuations did not reflect business reality per se but rather reality as filtered and defined by the measurement frameworks chosen by investors at that time. As market psychology shifted, so did the very framework by which companies were valued.

Practical Application: Investors should watch how technological uncertainty lets multiple valuation frameworks coexist for a time. Maintaining awareness of the metrics used to justify current valuations is crucial. This also requires awareness about how shifts in these frameworks might trigger market realignments.

This quantum perspective acknowledges that market psychology, like quantum states, can temporarily defy fundamental realities. It can create both extraordinary risks and opportunities for investors who maintain conceptual flexibility.

Housing market pre-2008

The Quantum State: The pre-2008 U.S. housing market maintained a superposition of sustainable growth and a speculative bubble. Psychological factors such as extrapolation bias and overconfidence allowed contradictory market interpretations to coexist until the financial crisis forced a resolution.

The Story: The evolution of the U.S. housing market between 2000 and 2008 provides a powerful demonstration of behavioral quantum finance. It showed how psychological biases can create market conditions that simultaneously reflect contradictory realities. During this period, nationwide home prices increased approximately 85% (Case-Shiller Index) before collapsing by over 30% in the subsequent crash. This extraordinary cycle was driven not simply by economic fundamentals but by behavioral factors that created quantum-like states of valuation.

The bubble's formation stemmed from a potent combination of genuine economic factors and psychological biases. On the fundamental side, low interest rates, financial deregulation, and financial innovation, particularly mortgage securitization, expanded credit availability. However, these conditions alone cannot explain the market's extreme movements. Behavioral factors like extrapolation bias, where people project recent trends indefinitely into the future, played a crucial role in sustaining price increases well beyond levels justified by fundamentals.

This extrapolation bias created a unique form of market superposition, where housing prices could simultaneously represent both investment returns and living expenses. Real estate investments were widely perceived as low risk during the bubble, even without any objective validation. This perception helped homebuyers justify purchases at increasingly unsustainable price levels. This created a market where prices could rise far beyond what income levels or rental yields would optimally support.

The psychological feedback loops supporting this superposition are particularly evident in markets with the most extreme price appreciation. In areas such as Las Vegas, Phoenix, and South Florida, speculative buying, involving purchasing homes primarily for anticipated price appreciation rather than occupancy, became commonplace. A 2006 National Association of Realtors survey found that 40% of home purchases were for investment or vacation properties rather than primary residences. A good 27% of all buyers were planning to resell within two years.

The role of overconfidence bias further reinforced this speculative psychology. There is clear historical evidence that real estate markets move in cycles. However, market participants increasingly viewed continued price appreciation as virtually eternal. This overconfidence extended to financial institutions. They relaxed lending standards, assuming that rising collateral values would compensate for borrower weaknesses. Subprime lending expanded dramatically, with subprime mortgages increasing from approximately 8% of originations in 2003 to over 20% by 2006. Exotic products like negative amortization loans and NINJA loans (No Income, No Job, No Assets) flourished. This allowed borrowers to purchase homes they would not have qualified to buy under traditional lending criteria.

Interestingly, even statistical models used to evaluate mortgage risk were using behavioral assumptions without explicitly acknowledging them. Credit-rating agencies assessing MBS relied heavily on recent default data. This implied an indefinite persistence of behavioral patterns from the housing market's upswing. The rating models used failed to capture the inherent risk of securities backed by subprime collateral.

The collapse of the bubble revealed how quickly market psychology could shift from optimism to pessimism. As defaults increased and home prices began to fall in 2007, the market's quantum-like superposition collapsed. The prices converged to fundamentals like affordability ratios and rental yields. By 2012, national home prices had fallen by approximately 33% from their 2006 peak, with much steeper declines in localized hot markets.

The Quantum Finance Lesson: The housing market prior to 2008 exemplified how behavioral biases can create market conditions similar to quantum states. Conflicting valuations coexisted in those states until a change in market psychology forced reconciliation. Just as a quantum system remains in a superposition until observed, housing prices existed both as assets for living and as vehicles for investment until a market shift required prices to align with financial fundamentals.

This parallels the quantum principle of complementarity, where a system can display seemingly opposite characteristics depending on the perspective from which it is measured. In the housing market, properties simultaneously served as places to live and as investment opportunities. When viewed as shelter, homes were evaluated against rental costs; when seen as investments, emphasis was placed on price growth. Throughout different stages of the market cycle, participants alternated focus between these perspectives, often overestimating their grasp of real estate trends while underestimating the influence of psychological factors in sustaining elevated prices.

Practical Application: Investors in real estate should consider how behavioral factors can temporarily maintain prices above levels backed by fundamental metrics. This requires being aware of current market narratives while consistently applying valuation frameworks based on income-to-price ratios, rental yields, and historical cyclical patterns.

Like quantum states, market psychology can maintain coherence despite contradicting fundamental realities. This requires valuation approaches that provide protection when these psychological states eventually collapse.

Looking ahead

Behavioral quantum finance reveals the profound truth that markets are not just mathematical systems but living ecosystems shaped by human consciousness and collective psychology. From FOMO-driven crypto bubbles to the quantum-like superposition of contradictory beliefs during the housing crisis, we've seen how psychological forces create market behaviors that defy traditional economic logic. Yet understanding these behavioral quantum effects is only part of the story. Markets don't exist in isolation; they operate within regulatory frameworks that themselves shape behavior and outcomes. Financial market policies create unique quantum effects, shaping the reality they regulate. Our exploration continues to the next chapter and examines how policymakers can design adaptive governance systems that acknowledge both the quantum nature of markets and the quantum psychology of their participants.

Chapter 13

Policy and quantum financial governance

"*The stones guide the river, yet the river shapes the stones*" goes an old Quantaland proverb.

Regulator Fatima was crestfallen. Her rigid crypto rules couldn't prevent the collapse of the NFT exchange that left investors exposed and markets reeling. Her regulations were unambiguous and stable, and these very inflexibilities of her policies led to the failure. This chapter revolutionizes financial regulation with quantum principles. It will expose you to adaptive frameworks and sandboxes. Examination of cases like Dodd-Frank and MiFID II will show how regulations reshape markets. This will teach and encourage you to design resilient policies that bend with shifting conditions.

The quantum mandate: rethinking the regulator's role

Old-school financial policymaking relies on the assumption of relatively stable systems that are manageable through comprehensive, fixed rules and static oversight. Quantum financial governance starts from a fundamentally different premise. It recognizes that financial systems exist in a state of constant flux and dynamic evolution.

This quantum perspective necessitates policies that are flexible and frameworks that change in tandem with the market. But this must be achieved without ever neglecting their core job of protection and stability.

The fluidity principle: adapting to a shifting market

Quantum financial markets demand regulatory frameworks that can adapt to changing conditions without losing effectiveness or creating dangerous gaps. This flexibility must carefully balance the competing demands of stability, innovation, predictability, and responsiveness.

DOI: 10.4324/9781003619635-13

Adaptive regulatory frameworks for emerging technologies

Financial technology creates quantum-like challenges that require regulations to evolve alongside technological development. Rather than crafting rules only after technologies mature, regulators should adopt technology-neutral frameworks focused on outcomes, ensuring consistent oversight regardless of the underlying technology.

The Payment Authority of Quantaland exemplifies this approach through their technology-neutral framework. They apply consistent consumer-protection standards to all payment systems, irrespective of being conventional, mobile, or blockchain-based. They adhere to the simple principle of same risk, same control.

Adaptive regulatory systems support a proactive approach that maintains financial system integrity while accommodating continuous technological advancement. Regulations evolve through practical experience rather than attempting comprehensive rules before understanding the technology. The Quantaland FinTech Authority uses quarterly reviews to adjust oversight based on market developments, enabling systematic adaptation without lengthy rulemaking processes.

Adaptive finance regulation explores and collects information through controlled experiments, adjusting regulations based on empirical findings. This dynamic approach addresses the inherent pacing problem between technological innovation and regulatory response. Sunset provisions set expiration dates for emerging regulations, forcing periodic reassessment to prevent outdated systems from persisting.

The quantum insight reveals that financial technologies exist in multiple regulatory states simultaneously, much like quantum superposition. Adaptive systems maintain this flexibility until market conditions reveal the optimal regulatory response.

Cross-border regulatory coordination challenges

Quantum financial markets create entanglement effects where regulatory actions in one jurisdiction instantly affect global markets. This requires sophisticated coordination mechanisms to retain focus.

Regulatory arbitrage emerges when participants shift activities between jurisdictions to exploit regulatory differences. This often undermines effectiveness and creates unfair competitive advantages. Quantaland faces significant pressure as the cryptocurrency rules of neighboring countries incentivize digital asset firms to relocate to more favorable regulatory environments. International standard harmonization seeks to reduce arbitrage through coordinated standards, but market evolution typically outpaces consensus-building among sovereign nations.

Data exchange agreements enable oversight of global institutions, but quantum entanglement means problems can spread faster than coordinated responses can be organized. Crisis coordination frameworks facilitate joint responses when quantum transitions threaten multiple jurisdictions, though they must carefully balance sovereignty concerns with collective-action needs.

Regulatory competition can foster effective rules that attract beneficial activity while maintaining strong oversight standards. The Digital Asset Regulatory Council of Quantaland coordinates with international partners to develop compatible frameworks while preserving appropriate domestic sovereignty.

The quantum principle demonstrates that regulation exists in a global superposition, where actions in one jurisdiction affect regulatory states worldwide. This reality calls for adaptive coordination rather than rigid harmonization.

Regulatory response speed versus market innovation pace

Quantum markets evolve faster than legacy regulatory processes can accommodate, creating dangerous gaps that threaten both innovation and stability. Regulatory lag occurs when rulemaking takes years to address innovations developed over months. The Securities Authority of Quantaland took three years to develop digital securities rules, during which the markets evolved so dramatically that the rules were nearly obsolete upon implementation.

Fast-track procedures enable rapid regulatory response while maintaining appropriate consultation and analysis. Emergency powers offer temporary authority to address immediate stability threats, with safeguards to prevent overreach while efforts to develop long-term solutions continue.

Industry consultative bodies use digital platforms and streamlined processes to gather stakeholder input faster than traditional methods. This helps them maintain comprehensive engagement despite accelerated timelines.

Regulatory pilot programs test oversight approaches before full implementation. They learn from market experience rather than blindly trusting theoretical models. Quantaland's FinTech Innovation Office runs fast-track consultations that develop guidance within 90 days for new technologies while maintaining rigorous engagement standards.

The quantum insight reveals that regulatory timing must balance immediate responses with thorough analysis, allowing market conditions to resolve this at an optimal speed.

Evolutionary regulation and learning systems

Quantum governance requires regulatory systems that learn and evolve from experience. They prove more effective than static approaches that regularly become outdated by market changes.

Regulatory feedback loops capture data on effectiveness and unintended consequences, enabling improvement based on observed outcomes rather than assumptions. Quantaland's Regulatory Analytics Division monitors market behavior and compliance patterns to identify where rules work effectively and where they need adjustment.

Evidence-based regulation uses data analysis and empirical research to guide decisions, avoiding reliance on theoretical models or outdated precedents in quantum markets. Regulatory trials test oversight approaches in limited contexts before broader implementation. This approach maintains learning effectiveness while minimizing negative impacts. Another important component of the life cycle is sunset-review processes. They force periodic reassessment of regulations to evaluate necessities and adjust them as market conditions evolve.

Stakeholder feedback systems gather ongoing input from regulated institutions on effectiveness and emerging challenges. Conventional monitoring systems often overlook this information. Quantaland's Consumer Finance Authority conducts annual consultations to gather feedback on regulatory burden and improve oversight approaches accordingly. The quantum principle suggests that good regulation adapts and changes based on market data rather than staying stuck in outdated methods.

Stakeholder engagement in dynamic regulatory environments

Quantum regulatory environments require stakeholder engagement processes that offer rapid input and feedback while ensuring comprehensive participation. Real-time consultation uses digital platforms to gather continuous stakeholder input, moving beyond periodic formal rulemaking. This enables regulators to track market developments and address concerns proactively.

Quantaland's Digital Authority allows ongoing input on regulatory issues, letting participants provide feedback before problems require formal responses. Industry working groups unite regulators and market participants to develop approaches for new technologies, enabling shared learning and more effective regulation.

Multi-stakeholder forums include consumers, industry representatives, and public groups in regulatory processes, ensuring all perspectives contribute to regulatory design. Transparent regulatory processes offer public access to analysis and decisions while appropriately maintaining confidentiality for sensitive information.

Regulatory impact analysis uses systematic cost-benefit reviews that consider effects on all stakeholders, ensuring consideration of the impact on consumers, institutions, and the economy.

The quantum insight reveals that effective engagement balances comprehensive participation with rapid response, allowing regulatory needs to be resolved through optimal consultation approaches.

Systemic risk monitoring

Quantum financial systems demand monitoring approaches capable of detecting emerging risks before they threaten systemic stability. These systems must account for the interconnected, rapidly evolving nature of modern financial markets and the quantum entanglement effects. We need to keep in mind that the entanglement can cause localized problems to spread instantly across the entire system.

Real-time risk detection systems and early warning indicators

Old-school risk monitoring, relying on periodic reports and historical data, often misses fast-evolving threats in quantum markets. Real-time systems identify emerging threats while intervention is still possible.

Continuous data monitoring processes market information constantly, enabling the detection of emerging threat patterns before they become systemic. Quantaland's Financial Stability Monitoring Center processes trading data, credit flows, and related indicators in real time, identifying unusual patterns that may signal systemic threats.

Early warning systems combine multiple data sources to spot conditions preceding financial disruption, flagging them for timely intervention. Pattern-recognition algorithms detect unusual market behaviors indicating emerging threats or manipulation, enabling regulatory responses before stability is compromised.

Threshold monitoring instantly alerts regulators to dangerous conditions when key indicators surpass predetermined levels. Market microstructure analysis tracks trading patterns to spot operational threats and manipulation that could affect market integrity.

The quantum principle reveals that risk monitoring balances emerging threats against system stability, with real-time measurement resolving uncertainty into specific states requiring intervention.

Network analysis for identifying systemic vulnerabilities

Financial networks create quantum entanglement effects where problems at single institutions can spark system-wide instability. Network analysis identifies these vulnerabilities before stress events test them.

Interconnection mapping identifies relationships among institutions that could spread stress, helping regulators understand potential contagion pathways. Quantaland Central Bank maps interbank connections, derivative exposures, and payment dependencies to track how stress might propagate through the system.

Concentration risk analysis identifies when multiple institutions face similar risks or rely on shared providers, creating hidden systemic vulnerabilities that traditional oversight might miss. Critical node identification pinpoints systemically important institutions and infrastructure for targeted oversight and contingency planning.

Cascade simulation modeling tests how problems at individual institutions might spread through networks, helping regulators design effective intervention strategies. Dynamic network analysis tracks evolving financial relationships, spotting emerging risks from new connections and exposures. Quantaland's Systemic Risk Assessment Team conducts quarterly network analysis, identifying shifting interconnection patterns and concentration risks that could threaten financial stability.

The quantum insight shows systemic risk exists in multiple network states simultaneously, with stress events collapsing these possibilities into actual contagion pathways.

AI and machine learning in regulatory surveillance

AI enables regulatory surveillance at scales and speeds that are beyond human capability. This ability makes using such technology essential for monitoring quantum markets with their vast, rapidly changing data flows.

Anomaly detection algorithms spot unusual patterns in market data that may signal emerging threats, manipulation, or operational issues requiring regulatory attention. Quantaland's Market Intelligence Unit uses AI to monitor millions of transactions daily, identifying trading patterns that indicate insider trading, market manipulation, or systemic threats. This required them to train their models with the ability to identify emerging patterns, too. Predictive threat modeling uses ML to identify institutions and markets with elevated risk levels.

Data are drawn from sources to achieve proactive oversight. Natural language processing examines communications, reports, and media coverage to spot emerging threats not evident in quantitative indicators alone. Behavioral pattern recognition detects changes in market participants' behaviors that may signal emerging threats or conditions requiring regulatory attention.

Automated report generation frees regulatory staff to focus on interpretation and response rather than data compilation. Quantaland's Regulatory Technology (RegTech) division uses AI not only to generate comprehensive threat reports for regulators. The report automatically highlights key issues with detailed analysis and recommended actions.

The quantum principle shows surveillance balances emerging threats against normal activity, with AI helping resolve complex data streams into specific threat patterns requiring action.

Cross-market correlation monitoring

Quantum financial markets exhibit correlation patterns that shift rapidly, often diverging from historical relationships. Cross-market monitoring identifies emerging patterns that could create systemic risks.

Asset class relationship tracking monitors correlations between stocks, bonds, commodities, and currencies. This process spots unusual patterns that may signal systemic risks or mounting stress. Quantaland Market Analytics tracks relationships across asset classes, alerting regulators when patterns deviate significantly from historical norms. The definition of significance is an issue that they decide in consultation with the regulators.

Geographic contagion monitoring tracks how financial stress spreads across countries and regions, enabling coordinated responses to potential crisis transmission. Similarly, analysis of sectoral risk transmission monitors how problems in one industry might spread across institutions and sectors, identifying emerging systemic risks.

Liquidity correlation assessment tracks simultaneous liquidity changes across multiple markets, identifying conditions that could create self-reinforcing liquidity problems. Volatility spillover analysis monitors how volatility in one market affects others, spotting stress-amplification patterns across interconnected systems.

The quantum insight reveals that cross-market relationships balance between independent and entangled states. Market events collapse these possibilities into specific contagion patterns that may require coordinated regulatory action.

Stress-testing evolution for quantum market conditions

Legacy stress testing relying on historical scenarios proves inadequate for quantum markets, which require testing for unknown scenarios and rapid correlation shifts. Stress tests based on innovative scenarios go beyond historical experience. It includes hypothetical extreme conditions that quantum markets might face.

Dynamic correlation testing reviews institution performance when asset correlations shift dramatically, reflecting the entanglement effects characteristic of quantum markets. Multiple-scenario testing evaluates performance across varied scenarios rather than single cases, providing a more profound understanding of resilience in quantum market conditions.

Real-time stress testing enables rapid assessment of institutional vulnerability during market stress, providing timely data for intervention decisions. Behavioral stress testing incorporates how management and participant behavior might shift during crises. This emphasizes an important feature: that human responses can amplify or mitigate effects. Quantaland's Insurance Stress Testing Program uses behavioral scenarios to test insurers' ability to handle unexpected claims and shifts in policyholder behavior during crises.

The quantum principle shows that stress scenarios balance multiple potential crisis states. Actual market conditions collapse these possibilities into tests that reveal specific institutional vulnerabilities.

Data-sharing challenges and privacy considerations

Effective systemic risk monitoring requires extensive data sharing between institutions and regulators, but quantum markets pose significant privacy and competitive challenges. Confidentiality-protection mechanisms enable information sharing for systemic risk monitoring while safeguarding sensitive competitive information and maintaining customer privacy.

Quantaland's Financial Data Authority operates secure platforms that allow institutions to share aggregate risk information after masking sensitive customer data or competitive details. Anonymization techniques enable the analysis of transaction patterns and risk exposures without identifying specific institutions or customers. This practice supports systemic monitoring while protecting privacy.

Regulatory information-access frameworks balance oversight needs with institutional concerns about protecting information of competitive value. Regulators provide clear guidelines on access and usage. International information-sharing agreements support cross-border risk monitoring while respecting varied national privacy laws and regulatory frameworks. Technology-infrastructure security protects shared information from cyber threats and unauthorized access, ensuring stakeholder confidence in data-sharing arrangements.

The quantum principle reveals that information sharing balances transparency against privacy concerns, with regulatory needs resolving this tension into mechanisms that ensure effective oversight while protecting confidentiality.

The quantum toolkit: principles for smarter regulation

Quantum approaches to financial regulation can significantly improve effectiveness by embracing uncertainty, enabling controlled experimentation, and adapting dynamically to changing conditions. This approach is fundamentally more effective than trying to control outcomes with rigid, detailed rules that will inevitably become outdated.

Probabilistic regulation: embracing uncertainty

Traditional regulation assumes deterministic relationships where specific rules produce predictable outcomes. Probabilistic regulation acknowledges inherent uncertainty and designs regulatory frameworks that work effectively across multiple potential scenarios.

Risk-based regulatory capital requirements

Risk capital requirements can utilize probability distributions instead of point estimates, creating adaptive frameworks that respond to evolving risks while ensuring systemic safety. Dynamic capital requirements change depending on real-time risk measurements rather than using static formulas. This approach optimizes efficiency while ensuring safety across different risk environments. Quantaland Banking Authority uses risk-based capital rules that increase during high-volatility periods and decrease when conditions stabilize, ensuring consistent safety across market cycles.

Scenario-based capital planning requires institutions to demonstrate capital adequacy across multiple stress scenarios, ensuring resilience in quantum market conditions, where correlations can shift rapidly. Institution-specific requirements consider individual risk profiles rather than applying uniform standards, optimizing efficiency while maintaining system-wide safety. Probability-weighted capital calculations use statistical distributions that account for various outcomes rather than focusing exclusively on worst-case scenarios.

Model validation ensures internal risk models remain relevant by regularly testing them against current market conditions rather than historical relationships.

Quantaland's Insurance Authorities require insurers to prove solvency across multiple economic scenarios, ensuring resilience in quantum conditions while optimizing capital efficiency.

The quantum principle reveals that optimal capital requirements balance efficiency and safety. Risk measurements resolve the fluid situation by balancing them at appropriate levels for current conditions.

Scenario-based policy frameworks

Regulatory policies can be designed to work effectively across multiple future conditions, avoiding the pitfall of optimizing for single scenarios that may not occur. Contingent policy responses provide preset rule adjustments that activate automatically under specific market conditions, enabling rapid adaptation without lengthy bureaucratic processes. Quantaland's Monetary Policy Framework uses contingent interest-rate policies that activate when inflation or unemployment exceeds defined thresholds, ensuring swift economic responses to changing conditions.

Multi-scenario impact analysis evaluates proposed rules across varied future conditions, ensuring effectiveness throughout quantum market evolution rather than for a single scenario. Adaptive rules trigger automatically, adjusting regulations when indicators show current rules have become unsuitable for the changed market conditions. Maintaining regulatory options involves keeping multiple potential responses available rather than committing to a single approach, enabling regulators to adapt rules as conditions and information evolve. Flexibility reserves allow emergency regulatory adjustments when quantum market transitions demand urgent responses beyond existing rule frameworks.

The quantum principle demonstrates that effective rule design balances multiple future states, allowing market conditions to resolve this balance into appropriate responses for current circumstances.

Uncertainty-aware regulatory impact assessment

Regulatory analysis can properly account for uncertainty by avoiding assumptions of precise outcomes and instead modeling ranges of potential impacts. Range-based impact estimates give us probability distributions for the costs and benefits of various rules, avoiding the false precision of single-point estimates.

Quantaland's monetary authority conducts regulatory impact studies that spread cost-benefit analysis across 25th, median, and 75th percentile outcomes. This enables informed decisions about uncertain rule impacts.

Sensitivity analysis examines how rule effectiveness varies under different market and economic assumptions, identifying robust policies that work across multiple scenarios. Assessment of unintended consequences proactively identifies potential negative effects that rules might create, recognizing that complex systems often respond in unexpected ways. Model uncertainty acknowledgment recognizes limitations in economic models used for rule analysis, providing appropriate caveats about prediction accuracy and underlying assumptions. Stakeholder impact distribution analysis examines how the costs and benefits of rules affect different groups, ensuring decisions account for distributional effects across market participants.

The quantum principle shows that rule impact exists as a balance between multiple outcome distributions. Implementation resolves uncertainties into actual costs and benefits for stakeholders, and that is when we understand the true impact of the regulation.

Dynamic regulatory thresholds and trigger mechanisms

Regulatory requirements can adjust automatically to changing market conditions, avoiding the problem of static thresholds becoming unsuitable for new environments. Setting up triggers for market conditions allows rules to change automatically when signs indicate that fixed rules no longer fit the current situation. Quantaland Securities Authority uses dynamic margin requirements that increase during high-volatility periods and decrease when markets stabilize, ensuring consistent investor protection across market cycles.

Institution-specific triggers account for individual risk profiles when defining appropriate regulatory responses, enabling tailored oversight for specific conditions. Cross-market coordination ensures trigger mechanisms work coherently across different regulatory domains, preventing conflicting requirements when multiple triggers activate simultaneously.

Predictive trigger indicators use leading rather than lagging indicators, enabling rule adjustments before problems emerge rather than after damage occurs. Override mechanisms are important too. It provides regulatory discretion to suspend automatic triggers when judgment indicates mechanical responses would be unsuitable for particular circumstances.

The quantum principle reveals that optimal rule thresholds balance static consistency with dynamic adaptation, with market measurements resolving this balance into appropriate responses for current conditions.

Outcome-based versus process-based regulatory approaches

Regulations can focus on achieving specific outcomes, allowing institutions flexibility in methods. This is achieved while avoiding mandated processes that become obsolete as conditions change. Performance standards set outcome objectives while enabling institutions to choose implementation methods, encouraging innovation while maintaining accountability.

The Quantaland Consumer Finance Authority needs certain consumer-protection focused results but lets companies choose how to achieve them, encouraging new ideas while maintaining protection standards.

Safe harbor provisions protect institutions from penalties when they reach non-conforming results using unconventional methods, promoting positive innovation. Outcome-focused reporting highlights results instead of how things are done, making it easier for institutions to follow rules. Incentives for innovation offer regulatory flexibility for institutions developing better methods to meet objectives and encourage continuous compliance improvement. Principle-based enforcement emphasizes outcome achievement over strict process compliance, supporting flexibility while ensuring regulatory objectives are met.

The quantum principle demonstrates that effective regulation balances rigid processes with flexible outcomes, with institutional performance resolving this balance into appropriate compliance methods.

Regulatory stress testing and contingency planning

Probabilistic regulation requires stress-testing frameworks that go beyond historical scenarios, testing resilience against quantum market conditions and unknown risks in complex systems. Quantum scenario development creates extreme tests accounting for nonlinear, interconnected markets, avoiding assumptions of gradual deterioration based on historical patterns.

Quantaland Central Bank uses quantum-inspired modeling for stress scenarios that include sudden correlation changes, liquidity evaporation, and behavioral shifts beyond usual recession-based scenarios. Multiple probability pathway testing examines how extreme factors interact to create systemic vulnerabilities. This approach acknowledges that quantum markets may experience several amplifying stresses at the same time.

Incorporating behavioral stress factors shows how management and participant behavior shift during crises, acknowledging that human responses can amplify effects through feedback loops. Financial models often overlook this factor. Cross-system stress testing evaluates how crises in one part of the financial system spread through interconnections that may only become apparent during extreme conditions.

When tests reveal vulnerabilities or actual conditions exceed normal response capabilities, contingency activation mechanisms quickly trigger preset regulatory responses. Recovery and resolution planning require systemically important institutions to develop credible wind-down processes that can be implemented without destabilizing the broader financial system. Quantaland Financial Stability Authority maintains contingency plans with measures including emergency liquidity provisions, temporary rule relaxation, and coordination protocols. All these can be activated within hours to address quantum market threats.

The quantum principle shows that systemic risk exists in superposition across multiple crisis pathways. The events collapse these possibilities into actual scenarios requiring coordinated regulatory responses.

Sandbox environments

Regulatory sandboxes enable controlled experimentation with new financial technologies and business models. These sandboxes allow innovation to flourish while maintaining essential consumer protection and systemic stability. These structured environments provide a crucial bridge between old-school regulatory approaches and the dynamic needs of quantum financial markets.

Controlled innovation testing frameworks

Testing frameworks provide structured environments where new financial technologies can be tested under relaxed regulatory rules while maintaining appropriate safeguards for consumers and market stability. These frameworks balance the need for innovation with the imperative of protection.

Limited-scope trials restrict activities to specific customer segments, transaction volumes, or business activities, enabling innovation while containing potential negative consequences. Quantaland Security Exchange's fintech sandbox limits participating companies to 1,000 customers and Q$1 million in transactions, ensuring that any problems that emerge don't threaten broader market stability.

Graduated scaling lets successful program participants gradually increase their activities based on demonstrated safety and effectiveness, creating a safe way to fully enter the market. Robust consumer-protection maintenance complements this approach. Participants have to uphold customer safeguards even under relaxed rules, thereby protecting stakeholder interests during innovation trials.

Exit procedures provide clear processes to end program participation when experiments fail or pose unacceptable risks, ensuring that failed trials don't threaten system stability or act as a disincentive to future innovation. Predefined metrics set objective criteria for evaluating program success. This also enables evidence-based decisions about which innovations should proceed with broader implementation.

The quantum principle shows that regulatory innovation balances controlled experimentation with full deployment, with testing results resolving this tension into clear pathways for successful technologies.

Regulatory safe harbors for emerging technologies

Safe harbor provisions protect innovators from regulatory enforcement as long as they stay within certain limits. This relaxation encourages helpful experimentation while keeping necessary rules in place. These protections create essential space for innovation to develop without premature regulatory constraints.

Technology-neutral protections focus on activities and outcomes rather than specific technologies, ensuring that safeguards apply consistently across various innovative approaches. Quantaland's Digital Payment safe harbor has temporary rules that ease regulations for new technologies for a short time, helping to avoid hindering innovation during important development stages. Good-faith compliance protection shields innovators from strict enforcement when they make reasonable efforts to meet regulatory requirements that may be unclear for new technologies.

Initiatives to improve regulatory clarity provide guidance on how existing rules apply to new technologies, reducing uncertainty that might otherwise discourage beneficial innovation. Industry consultation requirements ensure that protection provisions include stakeholder input before implementation. This supports innovation while incorporating risk-management guidance from experienced practitioners.

The quantum principle shows that innovation protection balances permissive experimentation with strict enforcement, with specific activities resolving this balance into appropriate regulatory coverage for emerging technologies.

Limited-scope trials and gradual scaling approaches

Innovation trials can begin with limited scope and expand gradually based on proven success, enabling learning while containing potential negative consequences. This phased approach allows regulators to gather evidence and build confidence in new technologies before permitting broader deployment.

Trial design creates structured environments with specific parameters for customers, transactions, and activities that can expand based on performance evidence. Quantaland's Blockchain Trial enables testing of distributed-ledger applications with set customer and transaction limits that increase based on proven security and effectiveness.

Risk-monitoring systems track trial performance continuously to identify problems before they threaten market stability, enabling early intervention when necessary. Performance benchmarks set clear criteria that trials must meet to proceed to the next level, ensuring that scaling decisions rely on objective evidence rather than subjective assessment.

Stakeholder feedback integration gathers input from consumers, industry participants, and other affected parties during trials, enabling continuous improvement and risk identification. Learning experiences are captured and analyzed to ensure that insights from trials influence broader policy development. This helps in maximizing the value derived from innovation experiments.

The quantum principle shows that innovation scaling balances limited trials with full deployment, with performance measurements resolving this tension into expansion pathways for successful technologies.

Cross-functional regulatory coordination in sandboxes

Sandbox programs require multiple oversight agencies to coordinate their efforts while maintaining their respective jurisdictions and expertise. This coordination ensures comprehensive supervision without creating regulatory conflicts or gaps.

Inter-agency coordination mechanisms ensure consistent oversight without conflicting requirements. Quantaland's Financial Innovation Council includes representatives from banking, securities, insurance, and consumer-protection agencies, providing integrated supervision for innovation testing.

Jurisdiction clarification determines primary supervision responsibility for different aspects of innovation, preventing regulatory gaps and interference. Information-sharing protocols enable regulators to share participant data while maintaining appropriate confidentiality, facilitating comprehensive oversight. Joint evaluation processes provide integrated assessment across different regulatory areas, ensuring comprehensive yet efficient supervision of innovative activities. This coordinated approach helps prevent regulatory arbitrage while ensuring that innovations receive appropriate oversight from all relevant perspectives.

The quantum principle shows that multi-agency coordination balances unified supervision with jurisdictional independence, with specific activities resolving this balance into appropriate coordinated responses.

International sandbox cooperation and standards

Cross-border financial innovation requires international cooperation in testing programs, enabling global technology trials while maintaining effective oversight coordination. This cooperation helps prevent regulatory fragmentation that could hinder beneficial innovation. Mutual recognition agreements allow program participants approved in one jurisdiction to operate in partner jurisdictions, facilitating cross-border testing. Quantaland maintains agreements with several countries that enable fintech firms approved in partner programs to test their services in Quantaland under streamlined processes.

Information-sharing arrangements facilitate coordination among international regulators for cross-border participants to avoid unnecessary duplication through comprehensive oversight. Common standards development creates compatible regulatory systems across jurisdictions, enabling testing scalability across borders without requiring separate approvals in each market.

Joint testing programs enable countries to collaborate on innovations that require cross-border operation such as international payment systems or global trading platforms. These cooperative efforts help ensure that regulatory frameworks evolve to accommodate increasingly global financial technologies.

The quantum principle shows that international testing cooperation balances national sovereignty with global coordination needs, with cross-border innovations resolving this tension into appropriate regulatory systems.

Success metrics and graduation criteria for sandbox participants

Clear criteria for program success enable objective evaluation of innovation experiments. This further creates evidence-based pathways to regulatory approval for full-scale rollout. These metrics provide transparency and certainty for innovators while maintaining regulatory standards.

Performance measurement frameworks set specific metrics that program participants must meet to demonstrate success, promoting evidence-based graduation decisions. Quantaland's Innovation Committee requires participants to prove compliance with consumer-protection laws, operational reliability, and financial sustainability before qualifying for full market operation.

Consumer outcome assessment checks whether innovations deliver genuine benefits without resorting to regulatory avoidance while meeting real market needs. It ensures that participants develop scalable capabilities appropriate for business growth beyond the sandbox environment.

Competitive impact analysis assesses whether innovations ensure fair competition, discouraging advantages gained through regulatory arbitrage rather than genuine innovation. Graduation pathway clarity offers transparent processes for moving from testing to full operation, giving innovators confidence while maintaining regulatory standards.

The quantum principle shows that program success balances innovation potential with compliance requirements, with performance measurements resolving this tension into graduation pathways for proven technologies.

Meanwhile, far from Quantaland

Cryptocurrency regulation post-FTX collapse (2022)

The Quantum State: Cryptocurrency regulation existed in a superposition of competing approaches. It ranged from light-touch innovation support to strict financial oversight. The FTX collapse forced policymakers to reconsider the balance between innovation and investor protection.

The Story: The November 2022 collapse of FTX, once the world's third-largest cryptocurrency exchange, represents a pivotal case study in quantum regulatory adaptation. Founded in 2019 by Sam Bankman-Fried (SBF), FTX grew rapidly to process over $10 billion in daily trading volume. It achieved a $32 billion valuation by early 2022. Its spectacular implosion was triggered primarily by revelations of misappropriated customer funds and flawed risk management. This forced regulatory authorities worldwide to reconsider their approaches to cryptocurrency oversight.

Before FTX's collapse, cryptocurrency regulation existed in a state of quantum-like superposition between competing frameworks. Different jurisdictions applied radically different approaches, from the innovation-friendly regimes of Singapore and Dubai to the more restrictive oversight models of China and India. Even within individual countries, competing regulatory perspectives created policy uncertainty. In the United States, the SEC advocated for treating many cryptocurrencies as securities, while the Commodity Futures Trading Commission (CFTC) argued for a commodities-based framework with lighter oversight.

FTX and SBF personally played a significant role in maintaining this regulatory uncertainty. They publicly advocated clearer regulations and even proposed a regulatory framework through the Digital Commodities Consumer Protection Act (DCCPA). At the same time, FTX exploited regulatory gaps by establishing its main operations in the Bahamas beyond the reach of major financial regulators. The industry actively sought jurisdictions with lax oversight. Whether intentional or emergent, regulatory arbitrage prevailed.

What made FTX's collapse particularly influential was SBF's carefully cultivated image as the responsible face of cryptocurrency. Prior to the implosion, SBF contributed around $40 million to political campaigns, gave testimony before Congress, and appeared in numerous positive media profiles. When this carefully constructed persona collapsed alongside FTX, it triggered a quantum-like shift in regulatory thinking.

The change in regulatory approach was both rapid and multidimensional. In the United States, the SEC intensified enforcement actions across the cryptocurrency sector. In Singapore, the Monetary Authority of Singapore (MAS) proposed new measures requiring cryptocurrency exchanges to better secure customer assets. The European Union accelerated the implementation of its Markets in Crypto-Assets (MiCA) regulation, while Japan introduced stricter rules on stablecoins.

Perhaps most telling was how regulatory thinking shifted. Initially, the regulators prioritized financial stability, even considering the potential of cryptocurrencies to disrupt legacy financial systems. The regulators then pivoted from stability to protection.

This regulatory quantum shift extended beyond government agencies to self-regulatory efforts within the cryptocurrency industry itself. Major exchanges such as Binance and Coinbase implemented "proof of reserves" systems to demonstrate custody of customer assets. Industry associations proposed new standards for exchange operations and disclosures.

The FTX collapse also fundamentally changed how policymakers conceptualized cryptocurrency risk. Before the failure, regulatory concerns primarily focused on unique risks inherent to the crypto ecosystem. They focused on technical vulnerabilities, blockchain-specific regulatory challenges, and the potential facilitation of illicit finance. FTX revealed that classical financial risks such as fraud, misappropriation of customer funds, conflicts of interest, and inadequate corporate governance remained equally important in this new technological context.

The Quantum Finance Lesson: The evolution of regulation after the FTX debacle serves as a clear example of quantum thinking in policy development. Major market events like the FTX collapse have the power to collapse a range of competing regulatory approaches into more defined frameworks. Much like quantum measurement, which turns a spread of possibilities into a specific outcome, FTX's downfall forced regulators to settle on more concrete oversight models.

This mirrors the quantum concept of decoherence, where systems move from a state of uncertainty to a definite outcome due to interactions with their environment. The FTX incident caused a kind of regulatory decoherence, pushing policymakers to recognize and respond to the tangible risks of uncertainty and motivating them to adopt clearer oversight positions.

The FTX episode also highlighted the difference between the theoretical promise of cryptocurrency and how it works in practice. This forced regulators to focus on actual market behaviors and risks rather than idealized versions of digital finance.

Practical Application: Finance professionals involved with policy creation should recognize how their regulatory frameworks fundamentally shape market behaviors. They should go beyond assessing how markets merely respond to the policies. This means developing adaptive regulatory approaches that acknowledge both the innovative potential of new financial technologies and the persistent human tendencies toward risk-taking and fraud.

This quantum perspective acknowledges that financial regulation, like quantum measurement, doesn't simply observe market realities. Instead, it actively participates in creating them through its measurement frameworks and enforcement priorities.

Implementation of the Dodd-Frank Act (2010)

The Quantum State: Financial regulation after the 2008 crisis existed in a superposition between principles-based and rules-based approaches. The Dodd-Frank implementation represented an ongoing effort to balance these complementary yet seemingly contradictory regulatory philosophies.

The Story: The Dodd-Frank Wall Street Reform and Consumer Protection Act was signed into law in July 2010. It represented one of the most ambitious attempts to reshape financial regulation in American history. Named after its congressional sponsors – Senator Christopher Dodd and Representative Barney Frank – the legislation emerged as a direct response to the 2008 financial crisis. It sought to address systemic risks through comprehensive regulatory reform. The act's implementation process over the subsequent decade offers a fascinating case study in quantum regulatory adaptation. It showed how regulatory frameworks continuously evolve through interaction between policymakers and market participants.

Over 2,300 pages, Dodd-Frank established a regulatory framework that simultaneously embraced both rules-based and principles-based approaches. This reflected a regulatory superposition that allowed adaptation to evolving market conditions. On the rules-based side, the act mandated specific requirements. This includes the Volcker Rule that prohibited proprietary trading by banks, minimum capital and liquidity standards, and stress-testing procedures. On the principles-based side, it established new regulatory bodies like the Financial Stability Oversight Council (FSOC) and the Consumer Financial Protection Bureau (CFPB). It had broad mandates to address emerging risks.

This regulatory duality created implementation challenges that required continuous adaptation. The objective was to improve financial stability while remaining concerned about balancing stability goals with industry costs.

The Volcker Rule implementation exemplifies how regulatory superposition evolved through market interaction. Initially, the rule proposed a straightforward ban on proprietary trading by banks. However, implementation of the rule required distinguishing between prohibited proprietary trading and permitted activities such as market-making and risk-hedging. Exceptions were allowed for cases such as market-making, hedging, treasury trades, and so on. In practice, these distinctions proved remarkably complex. The final rule, issued in December 2013 after three years of development, spanned 964 pages and included numerous exceptions and interpretive guidance.

Similarly, the creation of the CFPB demonstrated how regulatory frameworks shape market behaviors rather than merely responding to them. Endowed with broad authority over consumer financial products, the CFPB's very existence altered financial institutions' approaches to product development and marketing. The supervisors shifted from box-ticking to outcomes-based supervision. This enabled them to see how financial institutions treat the consumers beyond mere compliance.

Perhaps most telling was how Dodd-Frank implementation evolved through political transitions. When the Trump administration took office in 2017 with a deregulatory agenda, many expected a complete dismantling of Dodd-Frank protections. Instead, what emerged was a focused recalibration that maintained core provisions while adjusting implementation approaches and enforcement priorities. The Economic Growth, Regulatory Relief, and Consumer Protection Act of 2018 modified instead of repealing

Dodd-Frank. The thresholds for enhanced supervision were raised while the core regulatory architecture was maintained.

This regulatory change illustrated that rules-based and principles-based approaches, though different, can complement each other to enhance oversight. Seemingly contradictory regulatory approaches like rules-based versus principles-based and strict versus flexible represented complementary aspects of effective oversight. Regulation was acknowledged to require adaptation as implementation experiences provided new context.

The Quantum Lesson in Financial Regulation: The implementation of Dodd-Frank offers a vivid example of quantum thinking in regulatory governance. Rather than choosing one approach over another, effective oversight demanded a balance. It called for maintaining both precise rules and adaptable principles. This duality mirrors the behavior of quantum systems, which show both wave-like and particle-like qualities depending on how they are measured.

The concept of complementarity is at work here. By combining specific requirements with general principles, Dodd-Frank created a regulatory structure that could better withstand and adapt to the constant evolution of financial markets. This blend yields a more complete and resilient oversight than either a rigid or purely flexible framework could alone provide.

Ultimately, Dodd-Frank recognized that approaches often considered opposites, such as prescriptive rules versus adaptable principles and strict enforcement versus cooperative supervision, can work together as complementary truths. Their integration reflects a deeper understanding of what effective financial governance requires in an ever-changing market landscape.

Practical Application: Financial regulators should examine designing oversight frameworks that explicitly acknowledge measurement effects. They must recognize how regulatory approaches themselves shape market behaviors. This means developing regulations that balance precise requirements with adaptive capabilities. This would create systems that can evolve alongside the markets they govern.

The regulators need to accept that regulatory frameworks, however robust, would need dynamic calibration and appropriate adjustments. This quantum perspective acknowledges that financial regulation represents an ongoing process of interaction between observers and systems and not a fixed state of equilibrium.

EU's MiFID II directive

The Quantum State: European financial markets existed in a dynamic state between transparency and complexity. MiFID II represented an ambitious attempt to enhance market visibility without compromising efficient operation. MiFID II demonstrated a delicate regulatory balance, incorporating quantum-like measurement effects.

The Story: The implementation of the Markets in Financial Instruments Directive II (MiFID II) in January 2018 represented one of the most sweeping financial regulatory reforms in European history. Building on its predecessor MiFID I (implemented in 2007), this comprehensive regulatory framework had multiple objectives. It sought to increase transparency, strengthen investor protection, and improve market structure across European financial markets. The directive has over 400,000 words. Its development and implementation offer a fascinating case study in quantum regulatory effects. It shows how measurement frameworks fundamentally reshape the systems they observe.

MiFID II's core innovation was its unprecedented emphasis on market transparency across virtually all asset classes. The directive expanded pre-trade and post-trade transparency requirements beyond equities. It extended to bonds, derivatives, structured products, and even emission allowances. This revolutionary approach aimed to transform European financial markets from relatively opaque systems to highly visible mechanisms. It aimed to make price formation and transaction details readily apparent to all participants.

This transparency mandate created quantum-like measurement effects as it fundamentally altered the markets it sought to observe. Recording requirements fundamentally reshaped the basics of market operation. This recognition of regulatory measurement effects reflected quantum thinking, as it acknowledged that observation changes the system being observed.

The directive's implementation revealed how regulatory frameworks, like quantum measurements, can produce unintended consequences through their observation methods. One dramatic example was MiFID II's impact on research provision. The directive required unbundling research from execution services. This forced investment firms to price and pay explicitly for research. They could no longer obtain it "for free" through trading commissions. This seemingly technical change fundamentally altered the economics of research provision, leading to a significant reduction in coverage of smaller companies and emerging markets, which had lower trading volumes. This structural market change reflected the quantum principle that measurement frameworks don't simply observe reality but participate in creating it. The way research is produced and consumed has changed due to this regulation.

MiFID II also had a clock-synchronization requirement. It mandated trade timestamps accurate to within 100 microseconds for high-frequency traders, demonstrating how regulatory precision itself creates new market dynamics. This technical requirement forced extensive infrastructure investment across the industry and generated entirely new businesses focused on time-synchronization services. Microsecond timing altered execution strategies.

Perhaps most telling was how MiFID II triggered market adaptation through regulatory arbitrage. This was the financial equivalent of quantum tunneling, where particles pass through seemingly impenetrable barriers. Market participants developed new instrument types and trading venues specifically designed to bypass certain requirements of this comprehensive regulation. Dark pools, which had been limited by the directive, evolved into new forms. Systematic internalizers and periodic auctions maintained much of their functionality while technically complying with the new rules.

The directive's global impact further illustrated quantum-like entanglement effects in financial regulation. Although it is primarily a European framework, MiFID II has led to global market changes due to cross-border connections. U.S. firms trading with European counterparties need to comply with European standards, effectively extending the directive's reach far beyond EU borders. Like a powerful earthquake, the tremors were felt far away from the epicenter, forcing residents to take evasive action.

Even years after implementation, MiFID II's effects continue evolving through ongoing interaction between regulators and market participants. The European Commission's 2021 review acknowledged both intended and unintended consequences. As a result, they proposed adjustments to research unbundling rules and transparency requirements for certain instruments. Adjustments were made as regulators learned how the rules work in reality in the market.

The Quantum Finance Lesson: MiFID II highlights quantum effects in the creation of regulatory frameworks. It demonstrates that measurement systems do not just observe and record preexisting realities. Additionally, they actively transform the very systems they monitor. Much like the way a quantum observation turns probabilities into definite outcomes, regulatory frameworks shape and define market realities through their measurement tools.

This mirrors the quantum observer effect: the act of measuring a system unavoidably influences it. The transparency requirements set by MiFID II did not simply expose existing market behaviors. They contributed to fundamentally changing those behaviors. Observation itself became an agent of change, amplifying certain activities, curtailing others, and even ushering in entirely new practices that had not existed before.

Rather than merely unveiling a preexisting financial reality, MiFID II actively created a new market environment through its specific measurement techniques and regulatory requirements.

Practical Application: Financial regulators should recognize that their measurements do not simply observe market behavior; they also influence it. This means designing regulations with awareness of potential behavioral adaptations, unintended consequences, and systemic feedback effects.

To be effective, a rule must consider how it will interact with the market to produce real-world outcomes. This quantum perspective acknowledges that financial regulation, like quantum measurement, is an interactive process. It is not passive. The continuous dialogue between regulatory frameworks and market behaviors creates financial reality.

Looking ahead

Quantum financial governance represents a fundamental evolution in how we design and implement financial regulation, moving from rigid rules to adaptive frameworks that evolve with the markets they govern. Through regulatory sandboxes, probabilistic approaches, and outcome-based oversight, policymakers can create resilient systems that balance innovation with protection. Yet for all the sophisticated regulatory frameworks we develop, finance ultimately comes down to individual decisions about personal wealth and security. The quantum principles that transform corporate finance, investment management, and regulatory policy find their most personal application in how individuals navigate their own financial futures. Our next exploration reveals how quantum thinking can revolutionize personal wealth management, helping individuals create flexible, adaptive strategies that thrive regardless of life's uncertainties.

Chapter 14

Quantum finance for personal wealth management

"*Keep the fishes you catch close, but the baits closer*," goes an old Quantaland proverb.

Gig worker Liam had been a careful planner. He had planned a fixed budget for his freelance income and had always followed it. But a sudden inflation surge blew through his savings and left him scrambling. He had never realized that his discipline had eroded his adaptiveness. This chapter connects quantum finance to personal wealth and shows how to plan for multiple futures. The chapter examines FIRE and gig-economy cases and teaches you to build adaptable financial systems. It shows how to practice probability-based planning and create emergency funds that thrive in life's uncertainties.

The quantum mindset for personal finance

Personal finance exists in a state of quantum superposition where multiple potential life paths remain possible until specific decisions and external events collapse these possibilities into concrete realities. Understanding this fundamental quantum nature helps individuals build financial strategies that remain effective across multiple possible futures. This is an improvement over betting everything on a single predicted outcome.

This perspective transforms wealth management from an exercise in prediction to a process of preparation by creating plans that maintain flexibility and optionality no matter how life unfolds.

Planning for life's probabilities

Traditional financial planning assumes a predictable life progression where individuals can plan careers, health incidents, and family situations years in advance. Quantum life planning recognizes that uncertainty is fundamental. It builds financial flexibility that can adapt to whatever actually happens rather than what we predict will happen.

DOI: 10.4324/9781003619635-14

Moving from deterministic to probabilistic life planning

Classical financial planning makes detailed projections based on time-linked assumptions about career progression, investment returns, and life events that would rarely unfold as predicted. A probabilistic approach acknowledges this inherent uncertainty and creates strategies that remain effective across multiple possible scenarios.

Scenario-based life planning replaces single-path projections with multiple potential trajectories that account for varied career outcomes, health developments, and economic conditions. Rather than assuming steady career progression, quantum strategy considers job changes, entrepreneurship opportunities, and unexpected disruptions.

Senior professionals in Quantaland demonstrate probabilistic strategies by preparing for three distinct scenarios: steady employment continuation, transition to consulting work, and early retirement due to industry disruption. Instead of optimizing for a single outcome, their financial strategy functions effectively across all scenarios.

Probability-weighted financial targets assign likelihood estimates to objectives instead of treating all goals as equally probable. Flexible milestone strategies create adaptable timelines that can adjust based on life developments, avoiding rigid schedules that become obsolete when circumstances change.

Uncertainty reserves allocate resources for unknown future needs, often proving superior to trying to predict all expenses in advance. An average Quantaland family maintains uncertainty budgets representing 10% of income, covering unpredictable needs such as retraining, emergencies, and unexpected opportunities that detailed plans often miss.

The quantum principle shows that financial strategy exists in superposition across multiple life paths, with life events collapsing these possibilities into actual needs requiring adaptive responses.

Career uncertainty and multiple income pathway preparation

Modern professions exhibit quantum-like uncertainty, where traditional career-progression assumptions frequently prove incorrect. Quantum strategy prepares for multiple income sources and potential transitions rather than relying on a single employer-based career path.

Skill portfolio development creates multiple marketable capabilities instead of overspecializing in single roles that might become obsolete. Technology disruption and economic change make professional diversification essential for financial security. A Quantaland technologist maintains skills in software development, project management, and business analysis, creating multiple income pathways that provide security when any single skill faces market disruption.

Earnings diversification combines traditional employment with freelance work, investment income, and business ventures, avoiding overreliance on single sources that might vanish suddenly. Job transition planning prepares you financially for changes that might require additional education, temporarily reduced earnings, or geographic relocation.

Entrepreneurship optionality maintains the capability to start businesses or pursue self-employment when traditional employment becomes uncertain. Professional network maintenance creates relationship capital that offers opportunities during transitions, recognizing that career success often relies on network effects beyond technical skills alone.

The quantum principle shows that professional success balances multiple potential pathways, with market conditions collapsing these possibilities into actual opportunities requiring adaptive positioning.

Health and longevity planning across probability ranges

Medical outcomes exhibit quantum uncertainty, where genetic predispositions, lifestyle choices, and healthcare developments interact unpredictably. Quantum strategy prepares for multiple potential health scenarios rather than assuming predictable aging patterns and medical costs.

Longevity scenario planning prepares for a range of post-career periods depending on health outcomes and life expectancy improvements. Quantaland retirees plan for retirement scenarios ranging from 15 to 35 years, recognizing that medical advances and lifestyle factors could extend life beyond actuarial predictions.

Medical cost uncertainty budgeting prepares for expenses ranging from minimal to catastrophic, depending on unpredictable medical outcomes. Disability income protection helps manage the uncertainty of lost earnings due to health problems that could impact a person's ability to work, often covering cases that regular insurance doesn't include. Long-term care preparation plans for care needs that might never occur, occur briefly, or last for decades. Medical investment strategies manage today's healthcare costs while ensuring savings for the future. On the other hand, preventive investments offer choices for improved health and lower risks later on.

The quantum principle shows that medical planning exists in superposition across multiple potential outcomes, with medical events collapsing these possibilities into actual care needs requiring adaptive responses.

Retirement timing flexibility and scenario planning

Timing of retirement depends on uncertain factors, including health status, career satisfaction, economic conditions, and family responsibilities. All these make having a fixed retirement date increasingly unrealistic. Quantum strategy maintains flexibility about timing while preparing for multiple scenarios.

Flexible retirement planning prepares for retirement anywhere from age 55 to 75, depending on unpredictable circumstances. Quantaland Retirement Funds builds post-career savings suitable for various retirement ages, recognizing that optimal timing becomes clear only as the transition approaches.

Gradual retirement transition planning prepares for wind-down approaches involving reduced hours, consulting work, or career changes rather than sudden shifts to full retirement. Economic cycle timing considerations influence retirement plans by accounting for how market conditions affect optimal retirement timing, recognizing that economic downturns require different preparation than bullish markets.

Health-driven retirement scenarios prepare for timing influenced by health factors that might accelerate or delay retirement plans. Household responsibility adaptation accounts for timing changes based on family needs, such as elderly parent care or a spouse's health situations.

The quantum principle shows that optimal retirement timing balances multiple potential scenarios, with life circumstances collapsing these possibilities into actual decisions.

Education and family planning under uncertainty

Family strategy involves quantum uncertainty about children's educational needs, career paths, and geographic preferences that can make traditional education savings approaches inadequate for actual developments. Strategies to prepare for uncertain education costs may vary dramatically based on children's interests. These interests may range across technology-based learning and alternative credentials beyond traditional college assumptions. Parents in Quantaland save for education while recognizing that their children might pursue trade school, entrepreneurship, or other paths requiring different types of support.

Family size uncertainty affects financial strategy because actual family development might differ from expectations, requiring flexibility to adapt to real outcomes. A child development support strategy prepares for varied needs, including special education requirements, talent development opportunities, and health support needs that can't be predicted early in a child's life.

The quantum principle shows that family strategy exists in superposition across multiple development paths, with life events collapsing these possibilities into actual educational and support needs.

The impact of technology disruption on personal financial planning

Digital advancements create quantum uncertainty in personal finance. This happens as automation, AI, and digital transformation affect careers, investments, and financial services in ways that go beyond traditional planning approaches.

Automation and digital advancements affect industries on unpredictable time frames, often causing job displacements and career transformations that require financial preparedness. A worker in Quantaland invests in education and skills that can survive automation while building financial flexibility to handle job transitions when they occur.

Digital investment-opportunity evaluation considers how technological advancements create new options such as cryptocurrency and digital assets that traditional financial planning often overlooks. Privacy and security planning addresses financial risks arising from cyber threats that are created by increasing digital dependence.

The quantum principle shows that digital technology's impact exists in superposition across multiple disruption scenarios. Specific developments collapse these possibilities into actual career and investment changes that require adaptive responses.

Financial superposition: one is never alone

Personal financial goals can exist in quantum superposition, where multiple objectives remain simultaneously possible until specific life events and decisions collapse these possibilities into concrete priorities. Understanding financial superposition helps individuals avoid premature optimization that eliminates valuable flexibility and optionality.

Maintaining multiple financial goals simultaneously

Traditional financial planning often forces premature prioritization between competing objectives like retirement savings versus debt repayment. Quantum financial superposition maintains multiple priorities simultaneously until circumstances determine the optimal emphasis for current conditions.

The objective superposition strategy balances multiple financial priorities while avoiding rigid rankings that might prove incorrect if circumstances change. Quantaland professionals simultaneously save for retirement, early mortgage repayment, and education funds and maintain financial flexibility rather than optimizing any single objective at the expense of others.

Priority flexibility enables emphasis to shift with changing circumstances, avoiding rigid rankings that might become inappropriate over time. Resource-allocation superposition distributes funds across multiple objectives, preserving the capability to emphasize different goals as life circumstances evolve.

The quantum principle shows that financial objectives exist in superposition across multiple priorities until life events resolve these possibilities into actual resource-allocation decisions.

Balanced approaches to saving versus spending decisions

The traditional saving versus consumption trade-off assumes a zero-sum relationship where current consumption necessarily reduces future financial security. Quantum balance recognizes that optimal consumption patterns exist across probability distributions rather than at single optimization points.

Present versus future balance maintains quality of life today while building financial security for tomorrow, avoiding extreme compromises in either direction that create suboptimal outcomes. Quantaland families allocate resources for current enjoyment such as travel and entertainment, while maintaining robust retirement savings, recognizing that extreme saving or consumption strategies both create problems.

Experience-based investment strategies recognize that specific spending creates lasting value through skill development and relationship building. They ultimately enhance future earning capacity. Quality versus quantity consumption optimization considers purchases that provide lasting satisfaction rather than focusing solely on the lowest cost.

The quantum principle shows that optimal consumption balances present enjoyment with future security, with life circumstances resolving this balance into appropriate spending decisions.

Career and education investment superposition

Professional development investments exist in superposition, where education, skills, and networking create multiple potential career paths, avoiding premature optimization of single trajectories.

Education investment diversification creates multiple marketable skills rather than a single specialization that might become obsolete. This approach maintains professional flexibility across changing economic conditions. Quantaland students pursue technical and business degrees while developing creative skills, creating multiple professional pathways instead of narrow specialization.

Career development portfolios combine formal education, professional certifications, and experiential learning to create varied advancement opportunities. Network investment strategies allocate time to build and maintain relationships across industries rather than focusing narrowly on a single professional path.

The quantum principle shows that professional development balances multiple potential pathways. Market conditions resolve these possibilities into actual opportunities that require adaptive positioning.

Geographic and lifestyle flexibility preservation

Geographic and living choices create quantum entanglement where location, housing, and lifestyle decisions affect financial planning while influencing career options, costs, and family considerations.

Location independence reduces reliance on specific geographic areas for income and living satisfaction. This enhances flexibility to relocate based on opportunities. Quantaland workers develop location-independent income sources while maintaining professional networks that support relocation when circumstances change.

Housing flexibility strategies avoid overcommitment to specific housing assumptions, adapting as family needs and geographic preferences evolve. Living scalability maintains the capability to adjust living standards based on income changes, avoiding fixed expectations that require constant income levels.

The quantum principle shows that geographic choices balance multiple location and living options, with life events resolving these possibilities into actual relocation and housing decisions.

Debt versus investment decision superposition

Traditional financial advice often forces binary choices between debt repayment and investment, assuming optimal decisions are independent of future circumstances. Quantum superposition maintains both debt reduction and investment goals simultaneously.

Choosing between paying off debt and investing involves balancing repayment with the chance to grow wealth, so it's important not to focus solely on paying off debt or just on investing. Quantaland homeowners make extra mortgage payments while contributing to retirement accounts. This strategy reflects accepting that optimal allocation depends on unpredictable future interest rates and investment returns.

Liquidity preservation maintains accessible funds while servicing debt obligations. One remains careful to avoid over-optimizing debt reduction at the expense of emergency flexibility. Tax optimization balancing considers how debt interest deductions and investment tax benefits interact across different time horizons.

The quantum principle shows that debt and investment decisions balance reduction versus wealth accumulation, with financial conditions resolving this balance into optimal allocation strategies.

Risk tolerance across different life domains

Risk tolerance varies across the personal finance spectrum, where conservative approaches in some areas can enable more aggressive strategies in others. Quantum superposition maintains different risk profiles simultaneously rather than forcing uniform risk management.

Domain-specific risk application applies varied risk tolerances to career, investment, housing, and lifestyle decisions, accounting for how risk approaches interact across domains. Quantaland entrepreneurs take significant career risks through business ownership while maintaining conservative approaches to housing and emergency funds, creating balanced risk exposure across varied life domains.

Capacity versus tolerance balancing recognizes how financial capacity to handle risks differs from psychological comfort. This approach enables calibrated risk management across different situations. Time-horizon adjustment applies varied risk tolerances based on target time horizons rather than using a single predetermined risk profile.

The quantum principle shows that risk tolerance balances multiple domains simultaneously, with circumstances resolving this balance into appropriate risk-management strategies for current conditions.

The adaptive portfolio: budgeting for a changing life

Personal financial management requires dynamic approaches that can adapt to changing income, expenses, and life circumstances. This flexible methodology proves far more useful than static budgets and investment plans that assume predictable financial patterns, which rarely materialize in reality.

Navigating quantum leaps of life

Major life changes create quantum transitions where financial plans must adapt rapidly to entirely different circumstances. Understanding how to manage these transitions helps individuals maintain financial stability during periods of significant change and uncertainty.

Financial management for job transition and career change

Job transitions create financial uncertainty that requires adaptive management approaches, as normal financial patterns may not be maintainable during these abnormal circumstances. Income transition planning prepares for reduced or uncertain earnings during job searches, career changes, or industry shifts that might take longer than expected. Most employees in Quantaland build six months of living expenses before pursuing job transitions, enabling them to focus on finding suitable opportunities rather than accepting suboptimal solutions forced by financial pressure.

Skill investment timing balances education and training with income needs, recognizing that professional development often requires temporary earnings sacrifices. Network activation strategies use professional relationships to identify opportunities during transitions while maintaining long-term professional support systems. Management of benefits transition addresses healthcare, retirement, and other benefit needs during job changes, avoiding coverage gaps that could cause financial catastrophes.

The quantum principle shows that job transitions exist in superposition across multiple potential outcomes, with specific opportunities transforming these possibilities into actual income and employment changes.

Health crisis financial adaptation strategies

Health crises create quantum disruptions where normal financial planning becomes inadequate and requires emergency economic management that is essential for family survival. Medical expense management addresses immediate healthcare costs while preserving long-term economic stability, balancing necessary care with family security. Patients in Quantaland negotiate payment plans for major medical expenses while maintaining emergency reserves for living costs and additional medical needs. These negotiations may occur not only with medical service providers but also through financing agencies and other indirect channels.

Income replacement strategies address reduced earning capacity during illness, utilizing disability coverage, family support, and alternative income sources to maintain economic stability. Coverage coordination maximizes benefits from health and disability plans while minimizing out-of-pocket costs through careful planning. Care coordination strategy manages costs for ongoing medical care and potential long-term needs that extend beyond initial crisis periods.

The quantum principle shows that health crises balance recovery possibilities with ongoing medical needs, with specific outcomes resolving these into actual care requirements and economic demands.

Relationship changes and financial restructuring

Marriage, divorce, and relationship changes create economic entanglement effects where individual financial plans must integrate with or separate from shared obligations and opportunities. Asset division planning manages property and account division during relationship dissolution, preserving individual economic security while minimizing transaction costs. Quantaland divorce laws restructure shared assets, including real estate, retirement accounts, and investment portfolios, allowing individuals to maintain economic goals while meeting family support obligations.

Debt responsibility allocation addresses shared obligations during relationship changes, ensuring fair distribution while protecting individual credit and economic security. Income coordination planning manages spousal support, child support, and shared expenses, enabling both parties to rebuild economic stability. Lifestyle adjustment enables sustainable changes reflecting altered economic circumstances while maintaining a reasonable life quality during relationship transitions.

The quantum principle shows that relationship changes balance between shared and individual economic states, and legal actions resolve these into actual asset divisions and support obligations.

Geographic relocation financial planning

Geographic relocation creates comprehensive economic impacts, including housing cost changes, income adjustments, tax implications, and lifestyle modifications that require integrated financial planning. Cost-of-living adjustment enables economic maintenance despite changed expenses through budget adaptations and income optimization strategies. A relocator in Quantaland studies housing, transportation, and living expenses in target locations, negotiating income adjustments or identifying additional income sources to maintain economic stability.

A housing transition strategy manages buying and selling real estate, coordinating timing to minimize carrying costs and transaction expenses during relocation. Fiscal implication planning addresses tax differences between jurisdictions and property tax changes that might significantly affect net income. Professional impact assessment evaluates how relocation affects career prospects, income potential, and network development in new geographic markets.

The quantum principle shows that relocation balances multiple potential locations and outcomes, with moving decisions resulting in actual cost and income changes.

Unexpected windfall and inheritance management

While unexpected windfalls can lead to significant improvements in economic security, unsustainable lifestyle costs and poor decisions can squander once-in-a-lifetime opportunities.

A windfall integration strategy incorporates unexpected funds into long-term economic plans. This is very useful to avoid the temptation to treat these resources as spending opportunities without careful planning. In Quantaland, an inheritor typically allocates inheritance across debt reduction, investment, and measured lifestyle enhancement. They are cautious to maintain discipline to preserve long-term benefits from unexpected resources.

Tax optimization planning minimizes tax impact from inheritance, lottery winnings, or other gains through careful timing (if possible) and strategy. Lifestyle cost management focuses on maintaining reasonable spending levels despite increased economic resources. Investment diversification spreads unexpected funds across multiple investment types and time horizons, avoiding concentration on single investments.

The quantum principle shows that financial gains exist in a superposition between opportunity and potential waste, with allocation decisions resolving these possibilities into actual long-term economic impact.

Economic cycle adaptation for personal finances

Market cycles create systematic changes in employment, investment, and cost conditions that require adaptive financial management rather than static approaches that assume constant conditions.

Recession preparation builds economic resilience for reduced income and higher unemployment risk by enhancing emergency fund procurement and expense-reduction capabilities. Quantaland households maintain enhanced contingency funds during market expansion phases while reducing discretionary spending during recessions, adapting their financial management to cycle phases.

Inflation protection strategies adjust budgeting and investment approaches when rising prices erode purchasing power. These strategies focus on both income generation and expense management. Investment cycle timing adapts strategies to market cycle phases instead of maintaining static approaches that ignore cycle impacts. Employment cycle management recognizes how market cycles affect job security and advancement opportunities, adapting career strategy and financial planning accordingly.

The quantum principle shows that market adaptation exists in superposition across multiple cycle phases, with specific economic conditions resolving these possibilities into actual strategies appropriate for current circumstances.

Importance of emergency funds

Emergency funds represent quantum optionality that preserves financial flexibility during uncertain periods. Understanding emergency fund strategy helps individuals build financial resilience that enables adaptation to unknown future challenges that defy traditional prediction.

Quantum uncertainty and liquidity planning

Traditional emergency fund advice assumes predictable emergencies are manageable with a standard formula. Quantum uncertainty requires crisis strategies that account for unpredictable challenges that might occur simultaneously or in rapid succession. Multiple emergency scenario planning prepares for varied crises, including job loss, health issues, family emergencies, and economic disruptions that require different financial responses. Quantaland professionals maintain separate emergency funds for different crisis types, recognizing that various emergencies need varied liquidity levels and duration coverage.

Correlation risk consideration acknowledges that crises often cluster together, with job loss frequently coinciding with economic downturns while health issues align with increased medical expenses. Liquidity ladder strategies maintain emergency funds with varied accessibility levels, including immediate cash and longer-term accessible resources.

The quantum principle shows that emergency needs exist in superposition across multiple potential scenarios, with specific events collapsing these possibilities into actual liquidity requirements.

Emergency fund sizing for unknown unknowns

Traditional emergency fund sizing uses expense multiples that assume predictable costs. Quantum uncertainty requires the sizing of emergency funds to account for unknown unknowns that standard formulas cannot predict.

Expense multiple adaptation adjusts traditional emergency fund formulas based on individual risk factors such as job security, health status, family obligations, and economic conditions. The Quantaland Worker Welfare Society advises maintaining emergency funds covering six to nine months of expenses for workers with income volatility and health issues, creating higher requirements than traditional three-to-six-month recommendations.

Income replacement planning ensures emergency funds can replace income for extended periods rather than just covering brief crisis expenses. Cost escalation preparation accounts for how emergencies often increase costs beyond normal expenses, including additional medical, legal, or management costs.

The quantum principle shows that the optimal emergency fund size balances multiple potential magnitudes, with risk assessments collapsing these possibilities into appropriate sizing decisions.

Accessibility versus return optimization in emergency funds

Emergency fund management balances immediate accessibility with return optimization to ensure that funds are available when needed while preserving purchasing power over time. Liquidity tier structures maintain emergency funds across varied accessibility levels, including checking accounts, savings accounts, and short-term investments that offer different combinations of liquidity and returns. Quantaland Advisers suggest keeping one month of expenses in checking accounts, three months in high-yield savings, and six months in bond funds, achieving an optimal balance between access and return optimization.

Return preservation strategies protect the purchasing power of emergency funds through inflation-protected investments and money market funds while maintaining reasonable accessibility. Accessibility testing ensures that methods for accessing emergency funds remain functional during actual crises when banking systems might face disruption.

The quantum principle shows that emergency fund optimization balances maximum accessibility against optimal returns, with specific timing needs collapsing this balance into actual access requirements.

Multiple emergency fund strategies for different scenarios

Different emergencies require varied financial responses, suggesting that emergency fund strategies should prepare for diverse scenarios rather than uniform requirements. Crisis category planning maintains separate funds for health emergencies, job loss, family crises, and opportunity capture, recognizing that different situations require different amounts and accessibility levels.

Geographic crisis preparation considers location-specific issues like natural disasters that might require evacuation funds and relocation resources. Technology crisis planning prepares for cybersecurity breaches and identity theft issues that need immediate financial response and recovery capabilities.

The quantum principle shows that emergency scenarios exist in superposition across multiple types, with specific events collapsing these possibilities into actual fund allocation needs.

Behavioral aspects of emergency fund maintenance

Success of an emergency fund strategy depends on behavioral management that preserves fund integrity while maintaining motivation for building and sustaining these crucial resources. Emergency fund discipline maintains clear distinctions between emergency resources and other savings, preventing inappropriate use for non-emergency purposes. Quantaland employees use separate accounts at different financial institutions for emergencies versus regular savings, creating psychological barriers that help preserve fund integrity.

Fund-building motivation maintains enthusiasm for accumulating emergency resources by celebrating milestones and recognizing the security and peace of mind these funds provide. Strategies to resist temptations prevent inappropriate emergency fund use for wants disguised as needs. This calls for establishing clear guidelines for when fund utilization is truly justified.

The quantum principle shows that emergency fund discipline balances preservation needs against spending temptations, with actual spending decisions collapsing this balance into concrete utilization choices.

Meanwhile, far from Quantaland

Adjusting to inflation spikes (2022–2023)

The Quantum State: Individuals faced a superposition of transitory and persistent inflation scenarios between 2022–2023. This required simultaneous preparation for multiple potential inflation trajectories rather than a binary positioning.

The Story: The dramatic inflation surge of 2022–2023 presented individuals with one of the most challenging personal financial management environments in decades. Let us take the case of the United States. After years of subdued inflation averaging well below 2% annually, U.S. Consumer Price Index (CPI) inflation accelerated dramatically in 2021, reaching 7% by December and peaking at 9.1% in June 2022. This was the highest rate since 1981. This unexpected inflation spike, coupled with genuine uncertainty about its duration, created a quantum-like decision-making environment for personal financial planning.

What made this inflation episode particularly challenging was the genuine uncertainty surrounding its trajectory. Experts offered sharply divergent assessments, from those predicting brief, transitory inflation to those warning of prolonged 1970s-style stagflation. The range of possibilities was expansive.

This uncertainty created the need for quantum-like financial planning. One had to simultaneously prepare for multiple inflation scenarios rather than betting on specific outcomes. Sophisticated advisers helped clients maintain exposure to various inflation hedges while preserving core portfolio integrity. One had to acknowledge the impossibility of perfect inflation forecasting.

One illustrative approach came from Vanguard's Personal Advisor Services, which recommended maintaining a "barbell strategy" during this period. This involved simultaneously holding inflation-protected assets like Treasury Inflation-Protected Securities (TIPS) and I Bonds alongside growth investments that could potentially outpace inflation over longer horizons. The objective was to maintain long-term diversification while creating some inflation protection.

The inflation adjustment period also revealed how different demographic groups experienced fundamentally different inflation realities, reflecting another quantum-like aspect of this financial challenge. Data from the Bureau of Labor Statistics showed substantially different inflation impacts across age groups, income levels, and geographic regions. For seniors on fixed incomes, healthcare inflation at 10.7% (2022) was greater than general inflation and created disproportionate hardship. For young families, childcare costs rose particularly quickly at 13.8% (2022, BLS data). For rural residents, transportation expenses became especially burdensome as fuel prices spiked.

Individuals with different spending patterns were affected by inflation in different ways. This recognition of personalized inflation experiences reflected quantum thinking by acknowledging the probabilistic rather than deterministic nature of inflation's impact.

Perhaps most instructive was how the most successful individual responses involved continuous adaptation rather than single dramatic adjustments. One of the ways was to create predetermined decision points that would trigger specific actions if reached, allowing measured adaptation without overreaction. This approach avoided one sudden, big move and replaced it with a series of smaller responses to realized inflation data. This avoided the need to believe in a single forecast.

By early 2023, U.S. inflation had moderated to 6.4%, suggesting a potential peak but remaining well above the Federal Reserve's 2% target. This partial moderation vindicated neither extreme inflation scenario of a quick return to normal nor prolonged double-digit inflation. Instead, it highlighted the value of maintaining flexibility across multiple potential inflation trajectories.

The Quantum Finance Lesson: The inflationary period of 2022–2023 emphasized the value of quantum thinking in personal finance. Rather than relying on single-outcome predictions, effective planning involved preparing for several possible outcomes at once. Just as quantum particles exist within a range of probabilities, inflation, too, presents a spectrum of potential futures. Addressing this uncertainty calls for probabilistic planning instead of deterministic approaches.

This insight mirrors the quantum principle of superposition, where systems can exist in multiple potential states at once. The most effective responses to inflation involved maintaining a superposition of strategies. This involved hedging against continued high inflation while also allowing for gains if inflation subsided. This flexible approach stood in contrast to traditional methods that locked into a single prediction too early.

Ultimately, successful inflation strategies must recognize the inherent limits of forecasting. It is essential to accept that predictions will never be perfectly accurate and to build adaptability into financial planning. Overconfidence in any one approach is itself a risk, while humility and flexibility serve as the best safeguards.

Practical Application: Finance professionals like you should incorporate quantum thinking by designing financial plans that acknowledge fundamental uncertainty rather than relying on precise predictions. This means creating financial strategies with built-in adaptability, such as incorporating inflation-resistant elements while maintaining core financial discipline that serves well across multiple scenarios.

Long-term commitments based on short-term predictions are never wise. It is better to build resilience to multiple scenarios through diversification and continuous adjustment. This quantum perspective acknowledges that personal financial success, like quantum systems, involves probability management rather than prediction precision.

Gig-economy workers – the freelancers

The Quantum State: Gig workers exist in professional superposition between employment and entrepreneurship. They require financial strategies that simultaneously address the benefits and challenges of both states while preparing for rapid transitions between income states.

The Story: The dramatic expansion of the gig economy over the past decade has created a new class of workers. These workers exist in a quantum-like superposition between traditional employment and entrepreneurship. While definitions vary, approximately 36% of U.S. workers (59 million Americans) participated in the gig economy as of 2023, according to the Upwork "Freelance Forward" study. These workers range from rideshare drivers to freelance professionals to multi-platform creators. They all face unique financial challenges that defy conventional personal finance frameworks and require quantum-like adaptability.

What makes gig work particularly challenging from a financial planning perspective is its fundamental income uncertainty. Unlike traditional employment with predictable paychecks, gig workers often experience extreme income volatility. At times, their earnings are substantially above or below their average income in consecutive months. A JPMorgan Chase Institute study found that 55% of gig workers experienced more than 30% month-to-month income fluctuation, compared to just 20% of traditional employees.

This income volatility creates the need for quantum-like financial strategies that simultaneously address multiple potential income states. Ask any freelancer, and they will tell you that managing the period between money flows is more difficult than earning money. The real challenge lies in designing a system that would work under both situations.

To face this quantum-like income state, gig workers have developed innovative approaches. One of them is income smoothing, where gig workers immediately allocate incoming payments across multiple accounts serving distinct time frames and distinct purposes. These could be money for immediate living expenses, near-term expenses, a buffer for lean periods, long-term savings, and the like. Many Indian households still uphold an ancient tradition. The lady of the house would measure grains for cooking the supper and then remove a handful from there into a separate pot. This pot creates the reserve for a day when some guests drop in, or she has forgotten to restock. This creates resilience for fluctuations by acknowledging that they are inevitable; the only thing uncertain is when they happen.

The quantum nature of gig work extends beyond income volatility to benefit provision. Traditional employees typically receive benefits such as health insurance, retirement plans, and paid leave from employers. Gig workers must simultaneously function as both employer and employee, creating and funding their benefits. This dual role requires separating business and personal finances while recognizing their fundamental interconnection – the financial superposition.

Healthcare provision exemplifies this challenge. When Affordable Care Act insurance exchanges launched, they provided crucial health insurance access for gig workers. However, the process required complex income estimation, which was particularly challenging due to the variable earnings of gig workers. So the subsidy varied with income, and to be prepared for it, gig workers needed to maintain income projections at different levels. There would be best-case and worst-case scenarios for periodic income, and the healthcare budget should be built across this range and not optimized for a single income prediction.

Retirement planning presents similar quantum challenges. Traditional employees often benefit from employer-sponsored benefit plans with matching contributions and automated savings. Gig workers must create self-funded retirement structures through available vehicles. This requires discipline and complex contribution calculations, which are further complicated because of variable income. Some gig workers invest a fixed percentage of whatever they earn in such schemes instead of deciding on a fixed amount. Some even have a variable rate for such investments depending on the earning level achieved, reflecting a quantum-like adaptation. The system automatically adjusts contributions based on the current income status.

Perhaps most telling is how successful gig workers approach professional development. They simultaneously invest in both specialized skill enhancement and flexible capacity development. They need to have deep expertise in one area with broad capabilities across related domains. Chase Jarvis, a photographer and educator, referred to these as T-shaped skills. If the market changes, the gig worker can adapt without sacrificing the specialized skill that commands premium rates. During the COVID-19–triggered lockdown, professional trainers who had earlier experimented with web-based deliveries seamlessly migrated to online delivery. They could use advanced features during their training, which the new entrants were not even aware of.

The Quantum Finance Lesson: The financial management approach of the gig economy is a clear demonstration of quantum thinking applied to personal finance. Rather than following traditional, linear planning, gig workers often benefit from maintaining several financial systems simultaneously. Much like quantum particles that occupy multiple positions at once, gig-economy workers must ensure their financial strategies are robust enough to function across a variety of income scenarios and professional identities.

This perfectly aligns with the quantum concept of superposition, where systems can exist in multiple states at once. Effective gig worker financial strategies reflect this by balancing spending and saving, combining deep specialization with broad skill flexibility, and managing both immediate and long-term financial goals.

Uncertainty is ever-present, affecting not just the amount but also the timing of both income and expenses. This is reminiscent of quantum complementarity, where unpredictability in income and timing demands adaptable financial strategies that remain flexible on both fronts. With this approach, the need for precise forecasting is greatly reduced.

Practical Application: Gig workers and many freelance finance professionals have the same trait and should design financial systems that function across multiple potential income states rather than optimizing for single scenarios. This means creating deliberately redundant financial buffers, maintaining percentage-based rather than fixed-dollar financial allocations, and developing professional capabilities that balance specialization and adaptability.

Whosoever faces an uncertain future should be ready with plans for most of the scenarios that may evolve. This quantum perspective transforms gig-economy finance from a problem of prediction to one of probability management. This requires systems that maintain coherent financial superpositions across all plausible future states.

Early retirement movement (FIRE)

The Quantum State: FIRE practitioners exist in a superposition between working and retired states. They require financial strategies that simultaneously address both immediate financial independence and long-term sustainability across multiple potential future scenarios.

The Story: The Financial Independence, Retire Early (FIRE) movement represents one of the most interesting applications of quantum-like thinking to personal finance. Emerging primarily online in the 2010s, this grassroots financial philosophy emphasizes radical saving and investing to achieve financial independence well ahead of the traditional retirement age. FIRE, if you look from a quantum finance perspective, fundamentally reimagines retirement not as a binary state but as a probability distribution of financial autonomy.

At its core, FIRE challenges the classical notion of retirement as a discrete event where one transitions completely from working to not working at a predetermined age. Instead, it conceptualizes financial independence as a spectrum of freedom where individuals gradually increase their autonomy through deliberate financial choices. FIRE replaces binary retirement states with optionality, where work exists in superposition with leisure. It is the personal "measurement" (lifestyle choice) that collapses the waveform.

This reconceptualization creates a quantum-like superposition between working and retired states, requiring strategies that function effectively across multiple potential futures. FIRE practitioners typically target savings rates of 50–70% of income to achieve what they term as their "FIRE number." This is the investment portfolio size needed to support their expenses indefinitely based on a safe withdrawal rate. The conventional rate of savings is 10–15% of income.

The movement's mathematical foundation comes primarily from the "Trinity Study," which is officially titled "Retirement Savings: Choosing a Withdrawal Rate That Is Sustainable." This study analyzed how different portfolio withdrawal rates performed across historical market conditions. The study found that a 4% inflation-adjusted withdrawal rate was sufficient over a 30-year period, and it became known as the "4% rule." This ended up forming the basis of most FIRE calculations. Under this framework, financial independence requires accumulated investments equal to 25 times annual expenses, since 4% of this amount equals yearly spending.

What makes this approach quantum-like is its explicit probabilistic nature. The 4% rule may work well or fail completely depending on future market conditions. This acknowledgment of fundamental uncertainty represents quantum thinking applied to retirement planning. The FIRE movement evolved to include multiple variations that reflect the quantum-like recognition that financial independence exists as a spectrum rather than a binary state. Financial independence is an individual definition. One can decide where to belong within the spectrum between total autonomy and partial independence. Perhaps most instructive is how the community approaches sequence-of-returns risk. This is the vulnerability of retirement portfolios to poor market performance in early retirement years. Traditional retirement planning often addresses this risk through relatively simple "bond tents" involving temporarily increased bond allocations around the retirement date. FIRE practitioners have developed more sophisticated approaches that maintain quantum-like adaptability across multiple market scenarios.

One popular approach involves the creation of cash cushions, which calls for maintaining 2–3 years of expenses in cash equivalents while keeping the remainder in growth investments. This creates a quantum-like hedge that simultaneously preserves short-term spending capacity during market downturns while maintaining long-term growth potential. Another quantum strategy involves deliberately adjusting spending based on market conditions rather than maintaining fixed inflation-adjusted withdrawals. Agreeing to variable withdrawal increases the survival probability of the portfolio.

The FIRE movement's development mirrors the collapse of a probability wave into observable reality through several pivotal moments. The movement mathematically demonstrated how withdrawal rates exist as probability distributions rather than fixed rules. The movement reached critical coherence in 2018 when The New York Times article "Meet the People Who Retire in Their 30s" went viral, entangling FIRE principles with mass-market financial psychology. Like tiny particles that are observed at specific times, these milestones show when the complex ideas of financial independence turned into real cultural events. Each of these measurements, on its part, altered the system's future trajectory.

The Quantum Finance Lesson: The FIRE movement demonstrates quantum thinking in retirement planning by viewing financial independence as a spectrum of probabilities rather than as a simple, binary outcome. Instead of seeing retirement as a single, definite event, it treats financial freedom as a distribution of possibilities and develops strategies that remain resilient across differing future scenarios. Just as quantum particles exist in a range of possible states, financial autonomy, too, exists as a range of outcomes rather than one set point.

Whereas traditional retirement planning assumes a predictable, fixed path, the FIRE approach embraces uncertainty and probability. It favors strategies that adjust to changing circumstances, not those that rely on static assumptions. FIRE practitioners design financial systems to adapt and remain robust, even as the future becomes unpredictable. For example, maintaining a cash cushion alongside growth investments allows individuals to respond to market changes. This echoes quantum entanglement, where the state of one element determines how the other is used.

Adopting this quantum view shifts retirement planning from seeking certainty to managing probabilities. Like Schrödinger's cat, a FIRE portfolio simultaneously holds multiple potential outcomes until real-world events force a clear result. The true innovation of the FIRE movement is not simply early retirement but a conscious rejection of financial determinism in favor of proactive, probability-based planning.

Practical Application: Individuals should conceptualize retirement as a probability distribution of financial autonomy rather than a binary state. This means developing financial strategies with built-in adaptability involving variable withdrawal rates, multiple income streams, and deliberately conservative projections that acknowledge fundamental market uncertainty.

Proponents of FIRE look to maximize freedom while acknowledging uncertainty as a fundamental feature. This quantum perspective acknowledges that financial security, like quantum systems, involves effectively managing probability distributions rather than seeking illusory certainty in an inherently unpredictable world.

Looking ahead

Quantum finance transforms personal wealth management from rigid planning into dynamic adaptation, helping individuals navigate life's uncertainties with confidence and flexibility. From FIRE practitioners treating retirement as a probability distribution to gig workers maintaining financial superposition across multiple income streams, we've seen how quantum thinking creates more resilient personal financial strategies. Yet even as individuals apply quantum approaches to their personal finances, the broader financial landscape continues to evolve at an unprecedented pace. The convergence of AI, blockchain technology, and quantum computing is reshaping not just how we manage money but how financial systems themselves operate. Our final exploration examines how these emerging technologies will transform quantum finance from innovative theory into the foundation of tomorrow's financial reality.

Chapter 15

The future of quantum finance

"*No one fishes in the sea alone*," goes an old Quantaland proverb.

Community bank CEO Sofia struggled to integrate AI-driven analytics and was losing out to competitors who were outpacing her with blockchain and quantum tools. Old systems lagged in adapting to new realities. This chapter describes how AI, blockchain, and quantum computing are shaping the future of quantum finance. You will examine applications like DeFi and trading algorithms to gain a sense of the future. You will read about JPMorgan's innovations and appreciate the need for collaborative frameworks to implement these technologies for adaptive, resilient financial systems.

The technological convergence: building the future of finance

The convergence of AI, blockchain technology, and quantum computing is creating a new technological foundation that will fundamentally transform financial decision-making. These technologies enable unprecedented approaches to uncertainty management and pattern recognition that align perfectly with quantum finance principles. These will potentially revolutionize how we understand and interact with financial markets.

This technological convergence promises to make quantum finance's theoretical frameworks practically implementable. Moving from conceptual models to operational systems, these applications can navigate the inherent uncertainty and complexity of modern financial markets.

The AI power: making quantum-like financial decision

AI represents the first practical step toward truly quantum-inspired financial decision-making. This technology enables systems to process vast information networks, recognize complex patterns, and adapt to changing conditions in ways that fundamentally mirror quantum system behavior.

DOI: 10.4324/9781003619635-15

AI pattern recognition and market analysis

ML algorithms exhibit quantum-like traits by holding multiple hypotheses simultaneously until available data converges toward specific insights. This process mirrors quantum superposition, where multiple states exist until measurement collapses them into definite outcomes. Deep-learning detection identifies market relationships that traditional analysis often overlooks, revealing hidden correlations that evolve over time. Quantaland AI Systems uses neural networks to analyze millions of data points, identifying relationships that emerge across different market regimes and detecting correlation shifts during market transitions.

Natural language processing analyzes news, social media, and communications to identify sentiment trends that influence market behavior. They capture quantum effects where information measurement itself alters decision dynamics. Computer vision applications analyze satellite imagery and social activity to generate economic insights beyond traditional data sources.

AI analysis exists in superposition across multiple market relationships, with data measurements collapsing these possibilities into actionable trading insights.

Real-time adaptive trading systems

AI-powered trading systems adapt their approaches in real time based on changing market conditions. They embody quantum-like flexibility, where trading strategies exist in superposition until market conditions determine optimal approaches. Dynamic strategy selection enables trading systems to choose among multiple options based on current conditions, avoiding static methods that become ineffective when markets change. Quantaland Trading uses AI systems that maintain multiple trading approaches simultaneously, deploying the most effective strategy under current conditions.

Market regime detection algorithms identify when markets shift between behavioral patterns, enabling systems to adapt to fundamental changes in market structure. Exposure management adaptation adjusts position sizing and portfolio composition based on continuously evolving risk assessments.

Adaptive trading exists in superposition across multiple potential approaches, with market conditions collapsing these possibilities into optimal trading strategies.

Predictive modeling and scenario generation

AI systems generate multiple projection outcomes simultaneously, enabling quantum-like planning that prepares for varied futures rather than a single predicted outcome. Enhanced Monte Carlo simulation uses AI to improve projection generation, identifying realistic probability distributions beyond traditional models. Quantaland Trading generates thousands of market projections using ML trained on historical and current data, creating more accurate probability distributions.

Economic forecasting integration combines diverse data sources to generate projections that factor in complex interactions between economic variables. Stress scenario generation creates extreme projections that test resilience beyond historical experience, preparing for unprecedented market conditions. Predictive modeling exists in superposition across multiple potential projections, with market events collapsing these possibilities into actual outcomes that require adaptive responses.

Robo-advisory evolution

Automated investment guidance services are evolving to provide quantum-inspired planning that adapts to changing client circumstances and market conditions rather than relying on static allocation models. Dynamic goal management systems enable robo-advisers to help clients maintain multiple objectives simultaneously, adapting recommendations based on changing circumstances. Quantaland's robo-adviser adjusts portfolios based on life events, market conditions, and objective progress, avoiding static allocations that become suboptimal.

Behavioral learning integration provides automated support that helps clients make better decisions by recognizing and addressing behavioral biases. Tax optimization automation coordinates investment decisions with tax planning to maximize after-tax returns through continuous optimization.

Automated guidance exists in superposition across multiple client needs, with life events collapsing these possibilities into appropriate planning recommendations.

Fraud detection and risk management

AI systems provide sophisticated fraud identification and risk-management capabilities that extend to identifying threats and opportunities often missed by traditional rule-based systems. Anomaly detection algorithms spot unusual patterns in transactions and market activity that might indicate fraud or emerging risks.

Quantaland Security uses network analysis to identify relationships between accounts and entities that indicate coordinated illegal activity. This is also useful in identifying systemic risks that traditional monitoring misses. Behavioral biometrics analyzes user behavior patterns to identify suspicious activity based on deviations from normal actions.

Fraud detection exists in superposition across multiple threat patterns, with behavioral measurements collapsing these possibilities into security responses requiring immediate intervention.

Unchaining finance: blockchain and smart contracts

Blockchain technology enables new forms of financial instruments and transaction mechanisms that operate according to quantum-like principles of decentralization, transparency, and programmable automation. This technology creates systems where financial outcomes emerge from network interactions rather than from central control.

Decentralized finance evolution

Decentralized finance (DeFi) platforms exhibit quantum-like traits where economic services emerge from network effects rather than from central authority, creating new possibilities for innovation and risk management. Automated market-making uses algorithmic pricing that adjusts continuously based on supply and demand, creating quantum-like price discovery through network interactions. Quantaland's DeFi platform operates automated market makers that provide liquidity for digital assets, with pricing algorithms that adapt dynamically to changing market conditions.

Cryptocurrency owners can function as mini-banks and market makers by lending their holdings to provide liquidity and create new investment opportunities. Yield-farming strategies help investors get the best returns by shifting their assets between different DeFi platforms according to current profit opportunities.

Governance-token mechanisms enable community participation in protocol development. This creates democratic approaches to economic system governance that contrast with traditional centralized control.

DeFi platforms exist in superposition across multiple network states, with user interactions collapsing these possibilities into actual liquidity provision and pricing outcomes.

Programmable money and smart contracts

Smart contracts enable economic instruments that execute automatically based on predefined conditions, creating quantum-like financial products that adapt without human intervention. Conditional payment mechanisms enable automatic payment execution when specified conditions are met, significantly reducing counterparty risk. Quantaland's Smart Contract Platform allows businesses to create payment agreements that execute automatically based on delivery or performance milestones verified through data from business systems.

Insurance automation uses smart contracts to provide automatic payouts based on verifiable external data like weather measurements or flight delays. Escrow automation eliminates traditional third-party services by using smart contracts to hold funds until all transaction conditions are satisfied.

Smart contracts exist in superposition across multiple execution states, with contract performance benchmarks collapsing these possibilities into automatic execution.

Central bank digital currencies

Central bank digital currencies (CBDCs) will transform monetary policy transmission and financial system oversight, creating possibilities for real-time economic monitoring and policy implementation. Real-time monetary action enables central banks to implement policy changes that take effect immediately, creating quantum-like responsiveness in economic management. Quantaland Central Bank's digital currency enables real-time interest-rate adjustments that immediately affect all currency accounts.

Economic monitoring enhancement provides central banks with real-time data about economic activity through currency transaction patterns, offering unprecedented visibility into economic dynamics. Greater financial inclusion enables broader access to financial services through digital currencies that don't require traditional banking infrastructure.

CBDCs exist in superposition across multiple policy states, with economic conditions collapsing these possibilities into actual policy implementations.

Tokenization of assets

Asset tokenization allows people to own and trade small parts of assets that were hard to buy or sell before, opening up new ways to invest and making markets work. Real estate fractionalization enables ownership of properties through blockchain-based tokens that trade easily on secondary markets. Quantaland Digi Assets allow investors to own fractional interests in commercial real estate portfolios, offering lower minimum investments and significantly greater liquidity.

Art and collectible fractionalization create liquid markets for assets that were previously difficult to trade, democratizing access to alternative investments. Private equity fractionalization enables broader investor access through tokenized fund shares that can trade on secondary markets.

Asset fractionalization exists in superposition across multiple ownership structures. Market transactions collapse these possibilities into actual fractional ownership arrangements.

Regulatory technology integration

Blockchain technology enables new approaches to compliance and oversight, providing greater transparency while reducing costs for financial institutions and regulators. Automated compliance reporting uses blockchain transaction records to generate regulatory reports automatically, reducing compliance costs while providing more timely and accurate data to regulators. Quantaland's RegTech Platform helps organizations show they are following the rules by using immutable blockchain records, which keep things clear.

Real-time oversight enables regulators to monitor institutional activities instantly rather than relying on periodic reports, allowing for more proactive supervision. Cross-border compliance coordination uses blockchain to share information between regulatory agencies while maintaining appropriate confidentiality.

RegTech integration exists in superposition across multiple compliance states, with regulatory measurements collapsing these possibilities into actual verification responses.

The quantum leap: computing the impossible

Quantum computing will revolutionize financial analysis by enabling calculations that are currently impossible with legacy computers. This technological leap will particularly impact areas requiring optimization across large numbers of variables and complex probability calculations, fundamentally transforming how financial institutions approach problem-solving.

Quantum algorithms for optimization

Quantum computers excel at solving optimization problems involving large numbers of variables with complex interdependencies. This capability makes them ideally suited for financial applications such as portfolio optimization, risk management, and strategic planning, where classical computers reach their computational limits.

Portfolio optimization breakthroughs

Quantum algorithms can optimize portfolios across thousands of securities while accounting for complex correlations, transaction costs, and risk constraints that overwhelm classical optimization approaches. Multi-objective balancing models enable simultaneous optimization of return, risk, and liquidity objectives without forcing traditional trade-offs between competing goals.

Dynamic correlation modeling used by Quantaland Securities considers how correlations change over time rather than assuming static historical patterns. Real-time rebalancing enables continuous portfolio adjustment based on changing market conditions instead of relying on periodic calendar-based rebalancing.

Portfolio optimization exists in superposition across multiple potential configurations, with market measurements collapsing these possibilities into optimal allocations.

Risk scenario modeling

Quantum computers can simulate risk scenarios across millions of potential outcomes simultaneously, providing risk-management capabilities that far exceed classical computer limitations.

Monte Carlo acceleration techniques enable risk scenario simulation that would require years on classical computers, providing comprehensive analysis within practical time frames. Quantaland Risk generates millions of scenarios for stress testing that would be impossible for classical computers to process within reasonable constraints.

Tail-risk analysis simulates extreme scenarios that classical computers cannot handle due to computational complexity. Systemic risk simulation analyzes interconnections between institutions to identify vulnerabilities that classical computers cannot fully assess.

Risk simulation exists in superposition across multiple-scenario outcomes, with market events collapsing these possibilities into actual exposures requiring immediate management.

Trading strategy development

Quantum algorithms can develop and test trading strategies using multiple parameters and scenarios that classical computers cannot explore comprehensively due to computational constraints.

Parameter optimization explores millions of combinations to identify optimal configurations that classical computers cannot evaluate thoroughly. Quantaland Trading develops algorithmic strategies and tests parameter combinations across historical scenarios that would require decades of classical computing time.

Market microstructure analysis examines high-frequency trading interactions involving complex calculations beyond classical computing capabilities. Arbitrage opportunity detection searches for profit opportunities across multiple markets simultaneously, spotting advantages that classical analysis might miss.

Trading strategy development exists in superposition across multiple potential configurations, with market testing collapsing these possibilities into optimal approaches.

Regulatory stress testing

Quantum computers will enable regulatory stress testing that evaluates institutional performance across comprehensive scenarios, going beyond the current regulatory evaluation capabilities. Comprehensive scenario generation creates stress tests that account for complex interactions between economic variables and institutional characteristics.

Quantaland Market Regulators use quantum approaches to conduct real-time stress evaluation. This enables continuous assessment of institutional risk, avoiding the problem of periodic evaluations missing rapidly changing conditions. Cross-institution analysis assesses stress transmission between institutions through complex interdependencies that classical computers cannot adequately assess.

Regulatory stress testing exists in superposition across multiple crisis scenarios, with economic conditions collapsing these possibilities into actual results that require intervention.

Cryptography and security

Quantum computing poses a complex challenge for financial security. It renders many of our current cryptographic defenses obsolete while simultaneously empowering us to build more resilient systems. This isn't a simple technological upgrade. It demands a fundamental reimagining of how we protect financial data and transactions.

Quantum cryptography advantages

Quantum cryptography offers something that classical methods fundamentally cannot. Security in quantum cryptography is guaranteed by the laws of physics and not just mathematical complexity. This approach protects against threats from both classical and quantum computers. For instance, quantum key distribution uses quantum principles to create encryption keys, providing unbreakable security for sensitive financial communications. This is why institutions like Quantaland Advanced Security rely on it to safeguard their highest-value transactions.

The benefits extend further. Quantum systems can generate truly random numbers, a critical improvement over the pseudo-random algorithms used by classical computers. They also enable advanced digital signatures that are impossible to forge, even with quantum power.

Think of it this way. Until tested, a cryptographic system's security exists in the superposition of states across multiple potential attack vectors. It is only through active measurement and verification, which is analogous to quantum observation, that this potential collapses into a confirmed, secure reality.

Post-quantum cryptography transition

The financial world cannot wait for quantum attacks to become a reality. We must proactively migrate to quantum-resistant algorithms. This transition requires careful strategy. Method migration planning is essential for replacing current cryptographic standards with new, robust alternatives while ensuring systems remain compatible and operational.

The Quantaland Crypto Authority has adopted a phased approach, using hybrid security systems to combine classical and quantum-resistant algorithms. This protects against both present and future threats during the transition. One must also consider performance optimization, as some new algorithms require more processing power.

This entire period of change is a superposition of states. A system's future security exists in a flux of potential outcomes. The concrete implementation decisions we make are what finally collapse these possibilities into a definitive state of protection.

Quantum-safe financial infrastructure

Our entire financial infrastructure must evolve to be quantum-safe without compromising the speed and reliability the market depends on. This means upgrading payment-system protection with quantum-resistant encryption that can handle millions of transactions per second without introducing latency.

Similarly, trading system security must be fortified against quantum attacks while maintaining those crucial microsecond response times. The evolution of database safeguards is also critical; we must protect vast stores of financial data without crippling query performance.

Across all these applications, the infrastructure's protection exists in a superposition of readiness for various threat scenarios. The eventual emergence of a capable quantum computer will act as the moment of observation, collapsing this superposition and revealing whether our defenses meet the actual requirement.

Speculation on the evolution of financial decision-making models

Ultimately, the future of finance lies not in technology alone but in its fusion with human insight. The most powerful financial decision-making models will integrate these new technical capabilities with behavioral psychology and deep economic understanding. This holistic approach will create frameworks far more sophisticated and adaptive than any single discipline could achieve on its own.

Holistic approaches

The era of isolated, siloed expertise in finance is ending. The most effective decision-making frameworks of the future will be holistic, synthesizing behavioral psychology, economic theory, and technological capability into a unified whole. They will succeed because they account for the complex reality of markets. They will resemble deeply human systems operating within a technological ecosystem.

Behavioral-economic-technological integration

The next generation of models won't fight human nature; they will work with it. By merging behavioral insights with economic theory and powerful technology, these frameworks will leverage human intuition and creativity while using AI to guard against our well-documented biases and limitations.

Consider Quantaland Finance. Their platform doesn't replace human portfolio managers; it augments them. AI tools analyze patterns and flag potential biases in real time, while the humans focus on high-level strategy and nuanced client relationships. The result is a synergistic partnership that outperforms either humans or machines working alone.

The true breakthrough is in designing systems that learn from behavior rather than trying to eliminate it. Real-time feedback loops help decision-makers recognize a bias during its formation stage, preventing it from crystallizing into a costly error. Furthermore, during periods of high market stress, these systems can provide emotionally intelligent guidance, helping traders and managers maintain clarity when fear or greed would otherwise cloud their judgment.

Multi-stakeholder decision frameworks

The definition of value is expanding. Future financial decisions will systematically account for their impact on employees, customers, communities, and the environment, not just on shareholders. This isn't merely altruistic; it's a recognition that sustainable, long-term success depends on the health of this entire ecosystem.

Quantaland Finance exemplifies this shift. They use advanced AI modeling to visualize how a single corporate decision, such as a factory opening or a supply-chain change, ripples out to affect all stakeholders. At the same time, it tracks traditional financial performance. ESG factors are no longer a separate checklist but are hardwired into the economic calculus of every choice.

This integrated approach fosters long-term value creation by balancing immediate financial returns with the health of vital relationships and community standing. The outcome is a more resilient and sustainable business model that creates benefits for all parties involved.

Real-time adaptive planning

The static annual budget or five-year plan is becoming obsolete. In a world where markets and personal circumstances can shift in an instant, financial strategy must become a continuous, adaptive process. Real-time adaptive planning is about moving from a rigid blueprint to a dynamic flight plan that constantly adjusts to new conditions.

This means strategies can now dynamically adjust goals as opportunities and threats emerge, avoiding the trap of sticking to a plan that has clearly become irrelevant. Firms like Quantaland Consultants have moved to a model of continuous updates, ditching the annual review for a living strategy that evolves with changes in client needs and market developments.

This is powered by tools that monitor multiple potential futures at once, shifting strategic emphasis as certain scenarios become more probable. Trigger-based systems can make automatic changes when certain market conditions happen, allowing for quick responses while keeping human oversight for significant changes. This integrated, continuous risk monitoring replaces the dangerous lag of periodic reviews.

Quantum-inspired decision models

The most profound shift may come from models inspired by quantum principles themselves. Embracing uncertainty and superposition, these approaches enable strategists to simultaneously maintain multiple potential paths until a decision becomes necessary.

A superposition strategy, for instance, keeps several credible options actively available, avoiding the common pitfall of premature optimization that kills valuable flexibility.

These models also account for entanglement. They recognize that a decision in one area, like marketing spend, is deeply intertwined with outcomes in another, like supply-chain capacity. Furthermore, they focus on uncertainty optimization. Strategies are designed to perform robustly across a range of potential outcomes instead of betting everything on one predicted future. Finally, they practice observer effect management, carefully coordinating how and when they measure performance to avoid influencing the very outcomes they are trying to assess. This results in clearer insights and more effective strategies.

Interdisciplinary collaboration

The sheer complexity of future financial systems will make siloed expertise obsolete. Success will demand deep collaboration between finance professionals, technologists, behavioral scientists, and economists. Only by integrating these diverse perspectives can we develop solutions that surpass the limits of any single discipline.

Cross-field expertise requirements

The future finance expert must be a polymath, mastering traditional concepts while gaining fluency in psychology, technology, and complex systems. This goes beyond casual familiarity. Technical fluency means understanding the capabilities and limitations of AI, blockchain, and quantum computing well enough to evaluate applications and to collaborate effectively with specialists. Firms like Quantaland Professionals are already addressing this, providing technical education to finance experts to foster productive dialogue with developers and researchers.

This knowledge expansion must include behavioral science to understand cognitive biases and decision-making psychology, allowing for the design of systems that work with human nature. It requires systems thinking to grasp complex interdependencies rather than isolated parts. Working knowledge of data science is needed to interpret results with data teams, and regulatory-technology awareness is crucial to navigating how compliance evolves with technological change.

Academic-industry collaboration

Bridging the gap between theory and practice has never been more critical. Universities and financial institutions must collaborate closely to develop frameworks that address real-world quantum challenges.

This can take many forms. Research partnership programs unite academic theorists with industry practitioners to solve concrete problems.

The Quantaland University Partnership, for instance, combines theoretical research on quantum finance with direct application development. Curriculum-development integration is essential to ensure that finance education includes quantum thinking and teamwork skills. Faculty-exchange programs enrich academia with real-world experience and provide practitioners with a channel to contribute to research.

Joint funding initiatives target projects that offer both theoretical advancement and practical solutions. At the same time, student internship initiatives give emerging talent vital exposure and provide institutions access to a pipeline of qualified future professionals.

Technology-finance partnerships

No single institution can own all the necessary innovations. Financial firms will increasingly partner with technology companies to build the next generation of quantum-inspired applications.

Fintech partnerships merge profound industry knowledge with agile technological innovation. The Quantaland Tech partnership is a prime example, combining banking expertise with AI and blockchain development to create novel financial products. Open innovation platforms allow institutions to engage with multiple tech partners simultaneously, preventing the stagnation that can come from a single-vendor relationship.

Participating in regulatory sandboxes lets institutions and tech companies test new approaches in a controlled environment with oversight. Collaboration on standards development ensures new technologies integrate securely with existing infrastructure, and shared research initiatives help distribute costs and accelerate innovation timelines.

Global knowledge networks

The challenges of quantum finance transcend national borders, demanding global cooperation for sharing knowledge, harmonizing standards, and coordinating responses.

International research consortia pool diverse perspectives and resources to tackle problems too large for any one country. Parallel standards harmonization efforts are needed to ensure technologies function across borders while respecting different regulatory regimes. Sharing best practices allows countries and institutions to learn from each other's successes and failures, avoiding costly duplication. Establishing crisis coordination frameworks enables a swift, international response to future financial disruptions, and education exchanges ensure professionals develop a global perspective on quantum finance applications.

The future of finance will emerge from the intersection of technology, behavioral understanding, and economic insight. As AI, blockchain, and quantum computing mature, they enable new decision-making approaches grounded in innovative principles. Navigating this transformation will require interdisciplinary cooperation and adaptive thinking. A firm grasp of these core principles provides the foundation for maintaining focus on the ultimate goals of value creation, risk management, and human development.

The quantum future promises economic frameworks that are not only more effective and responsive but also more humane, better serving both individual and collective welfare.

Meanwhile, far from Quantaland

Quantum-inspired algorithms in trading

The Quantum State: Trading algorithms are evolving beyond classical computation to incorporate quantum-inspired approaches. These algorithms simultaneously evaluate multiple potential market states, creating more adaptive systems for navigating uncertainty.

The Story: The intersection of quantum computing concepts and algorithmic trading represents one of the most intriguing frontiers in finance. Although fully functional quantum computers for financial applications are still in development, inspiring stories exist about firms that tackle market uncertainty by employing quantum-inspired algorithms on legacy systems. These methods use ideas from quantum computing in trading strategies, allowing systems to consider several possible market outcomes at the same time instead of focusing on just one.

The evolution of quantum-inspired trading began in earnest around 2017. Firms like JPMorgan Chase and Goldman Sachs established research teams focused on quantum computing. Initially, these efforts concentrated on long-term quantum hardware applications, but they swiftly produced insights applicable to classical computing systems. They found significant benefits that quantum-inspired algorithms running on classical hardware can deliver.

One prominent application of quantum-inspired algorithms is in portfolio management. Traditional portfolio optimization faces computational challenges when dealing with large asset universes, complex constraints, and nonlinear relationships. Quantum-inspired approaches transform these problems into quantum formulations. These can then identify optimal portfolios across multiple simultaneous objectives such as maximizing expected return, minimizing various risk measures, and satisfying complex constraints. Quantum-inspired portfolio optimization models do not pick a single efficient portfolio. Instead, the model maintains a portfolio superposition across multiple concurrent market regimes.

Another significant application involves quantum-inspired Monte Carlo simulations for derivatives pricing and risk management. Traditional Monte Carlo methods face challenges when modeling complex derivatives with multiple underlying assets and path dependencies. Quantum-inspired approaches use techniques adapted from quantum amplitude estimation to achieve quadratic speedups in simulation convergence, enabling more sophisticated risk modeling. These algorithms achieve the same results as the traditional system but with a much smaller number of simulations. The advantage lies not in clock speed but in probabilistic representation. What changes is how uncertainty is incorporated in the model.

Perhaps the most inspiring development is how quantum-inspired algorithms are transforming market microstructure analysis. Traditional market-making algorithms typically classify market states into discrete categories and optimize behavior for each state independently. Quantum-inspired approaches maintain superposition of multiple potential market states, continuously updating their probability weights based on incoming data.

The quantum-inspired approach extends beyond individual trading strategies to entire market frameworks. Researchers at the Santa Fe Institute have developed quantum-inspired agent-based models. These models mimic financial markets as if they were quantum systems, where the actions of participants are uncertain until they interact with the market. Researchers have found these models remarkably effective at reproducing empirical market phenomena such as fat-tailed return distributions and volatility clustering, which classical models struggle to generate.

While fully functional quantum computers remain some years from mainstream financial applications, hybrid approaches that combine classical systems with early quantum processors are already emerging. In 2020, HSBC partnered with the quantum computing company PASQAL to develop hybrid classical-quantum algorithms for asset pricing and portfolio optimization. Similarly, Barclays and IBM announced a partnership focused on developing quantum algorithms for securities settlement optimization that could dramatically reduce capital requirements.

The Quantum Finance Lesson: Quantum-inspired trading shows that quantum principles can significantly improve financial decision-making, even without the use of quantum hardware. The key lies in rethinking how we approach uncertainty. Instead of simply increasing computational speed, these methods reframe markets as probability distributions, much like quantum physics treats particles. In this framework, market conditions are evaluated continuously as ranges of probability rather than fixed states.

This approach mirrors the quantum concept of superposition, where a system can exist in many possible states at once. Quantum-inspired trading algorithms adopt this principle by keeping multiple market models, strategies, and scenarios active simultaneously. They constantly update the probability assigned to each instead of making a single binary choice. Additionally, these algorithms recognize that their actions influence the markets they study, creating feedback loops that require ongoing adaptation rather than fixed, one-time optimization.

Practical Application: Trading systems used by financial institutions should incorporate quantum thinking by explicitly maintaining multiple simultaneous market interpretations rather than selecting a single model. This means designing algorithms that adjust probability weights across competing theories rather than choosing between them. This would lead to the creation of systems with inherent adaptability to regime changes.

Even before quantum computing becomes mainstream, one can make conceptual shifts toward probabilistic rather than deterministic modeling. This approach can yield immediate benefits, even when using legacy hardware. This quantum perspective acknowledges that financial markets, like quantum systems, behave as probability fields rather than deterministic mechanisms. Quantum trading doesn't predict the future. Instead, it maintains coherent adaptability across possible futures until markets force measurement.

Sustainability and ESG investing

The Quantum State: ESG investing exists in a superposition between values-based and value-based approaches, the former focusing on ethical and moral principles while the latter considers ESG as material risks and opportunities. The sustainable finance industry is increasingly integrating these seemingly contradictory perspectives into unified investment frameworks that acknowledge their fundamental interconnection.

The Story: ESG investing has undergone a quantum-like transformation in modern finance, evolving from simple ethical screening to a sophisticated approach that merges sustainability with financial performance. This dual focus challenges traditional binary thinking about impact versus returns.

The shift becomes clear when tracing ESG's roots. Early socially responsible investing (SRI) relied on negative screening and blacklisted tobacco, weapons, and other "unethical" sectors. While straightforward, this binary approach offered limited investment flexibility.

Today's ESG framework recognizes the deep interconnection between environmental, social, and financial factors. As BlackRock's Larry Fink noted, climate risk is investment risk, and sustainable strategies often deliver superior risk-adjusted returns. This reflects a fundamental shift where financial and sustainability considerations aren't competing priorities but complementary aspects of sound investing. This quantum-complementary approach has created remarkable market growth. Global sustainable investments reached $35.3 trillion in 2020, representing 36% of professionally managed assets, according to the Global Sustainable Investment Alliance. Perhaps more telling is how mainstream financial institutions have embraced ESG principles. About 98% of U.S. asset managers are either implementing or considering implementing ESG strategies, according to a 2022 Callan Institute survey.

Present-day ESG stands out through its materiality focus. Instead of using universally applicable metrics, it prioritizes sustainability factors that significantly impact financial outcomes in specific sectors. Sustainability Accounting Standards Board's (SASB) industry frameworks (2018) exemplify this, pinpointing key ESG issues across 77 sectors. As SASB founder Jean Rogers notes, materiality works probabilistically, causing ESG factors to gain or lose financial relevance across industries and time frames. This mirrors quantum logic, treating sustainability-finance links as dynamic probabilities rather than fixed rules.

The ESG movement's evolution reveals three quantum leaps: The 2006 Principles for Responsible Investment (PRI) launch marked finance's first coherent sustainability framework. In 2020, the SASB/IIRC merger created a professional superposition – blending materiality metrics with integrated thinking. Finally, the 2023 ISSB standards collapsed competing approaches into a unified global reporting system while preserving necessary flexibility through sector-based implementations.

The quantum analogy extends to ESG analytics. While traditional ratings relied on static company disclosures, AI tools now process real-time data such as news, social media, and filings, creating living ESG assessments that evolve with new information. This probabilistic approach introduces quantum-inspired vulnerabilities. Just as measurement affects quantum systems, ESG's reliance on dynamic data creates "greenwashing interference patterns," where bad actors manipulate disclosures to collapse ratings favorably. The 2022 SEC crackdown on ESG funds revealed how 43% of sustainability claims existed in superpositions between truth and misrepresentation until regulatory observation forced collapse.

The most significant evolution is the shift to transition finance, where institutions support decarbonizing companies rather than just excluding polluters. GFANZ, a $130 trillion coalition formed at COP26, embodies this approach by funding emissions reductions across all sectors. The approach believes that climate solutions require financing transitions, not abandonment. This forward-looking mindset treats companies as evolving systems, applying quantum-inspired probabilistic thinking to sustainable finance.

Diverging regulatory frameworks create geopolitical superposition. The EU's Sustainable Finance Disclosure Regulation imposes strict disclosure eigenstates, while the United States' principles-based approach maintains probabilistic compliance.

The Quantum Finance Lesson: ESG investing serves as a practical example of quantum complementarity in financial analysis. Here, values-based and value-based perspectives, once considered opposites, are now recognized as complementary frameworks. Much like quantum phenomena, where entities display both wave and particle characteristics depending on observation, sustainable investments can simultaneously promote both sustainability goals and financial returns, depending on the analytical lens applied.

This idea closely reflects the quantum principle of complementarity, where seemingly opposing views actually combine to offer a fuller understanding than either can achieve alone. In contemporary ESG frameworks, environmental, social, and financial factors are not isolated considerations but intertwined components of thorough risk and opportunity analysis. ESG investing shows that both impact-driven and return-driven strategies hold valuable insights, and together they form a more robust investment framework than what would be possible individually.

In this context, ESG factors can be viewed as call options that provide potential benefits in future sustainability transitions. For example, when a mining company invests in decarbonization today, it is effectively paying a premium for the possibility of future regulatory or social advantages, with the true value of those efforts determined by the changing probabilities of materiality over time.

Practical Application: Effective ESG implementation requires maintaining Schrödinger's capitalism, where profit and purpose coexist in superposition until they are measured against specific stakeholder expectations. Investors should recognize that sustainability and financial performance represent complementary rather than competing objectives. This means developing investment approaches that explicitly integrate material ESG factors into financial analysis rather than treating them as separate considerations.

Sustainability investing creates long-term value and is not about sacrificing returns for values. Quantum investing reveals that value creation requires addressing the full spectrum of risks and opportunities, including environmental and social. This quantum perspective acknowledges that investment success, like quantum reality, requires embracing complementary frameworks rather than forcing false dichotomies between financial and sustainability considerations.

Education for future financial professionals

The Quantum State: Financial education exists in a transitional state between traditional discipline-specific training and emerging interdisciplinary approaches. This aims to prepare professionals for a financial landscape characterized by increasing complexity, uncertainty, and interconnection.

The Story: The evolution of financial education for future professionals represents a fascinating example of quantum thinking applied to knowledge development. As financial systems grow increasingly complex, interdependent, and technology-driven, leading educational institutions are beginning to transform how they prepare students. They are moving from traditional, siloed approaches toward frameworks that emphasize interdisciplinary thinking, comfort with fundamental uncertainty, and a systems-level understanding.

While dramatic curriculum overhauls specifically focused on the way we have treated quantum finance are limited, several genuine trends are emerging in financial education. We have identified a few trends, which are classified according to their focus areas.

Technology Integration in Finance Programs: MIT Sloan's master of finance program has integrated ML and data science courses as core requirements since 2018. Stanford's Management Science & Engineering program includes computational finance tracks that combine traditional finance with computer science. Carnegie Mellon's computational finance program represents one of the longest running interdisciplinary approaches, combining finance, mathematics, and computer science since 1994.

Behavioral Finance Integration: The University of Chicago Booth School has expanded its behavioral economics offerings, building on Nobel laureate Richard Thaler's work. Duke's Fuqua School integrates behavioral insights throughout its finance curriculum rather than treating it as a separate subject. London Business School's behavioral finance courses examine how psychological factors affect market dynamics.

Complexity and Systems Thinking: The Santa Fe Institute offers complexity science programs that attract finance professionals seeking to understand market dynamics through complex adaptive systems. INSEAD's Advanced Management Program includes modules on complexity leadership and systems thinking. Wharton's executive education programs increasingly incorporate network theory and systems dynamics.

Industry-academia partnerships have also created opportunities for finance professionals to acquire skills that promote quantum thinking. The Goldman Sachs engineering campus was launched in 2020, focusing on technology skills for finance professionals. The course includes data science, ML, and software engineering alongside financial knowledge.

JPMorgan Chase's "Code for Good" annual hackathon connects students with real financial technology challenges and partners with universities to help develop fintech solutions for social good. Similarly, BlackRock's Aladdin certification training collaborates with universities to provide students with access to institutional-grade risk-management technology.

Many universities have included simulation-based learning in their finance curriculum. The simulations cover a wide range of areas, including trading floor simulations to teach market dynamics using real market data.

The Quantum Finance Lesson: The way financial education is evolving clearly demonstrates the influence of quantum thinking on knowledge development. Rather than relying on a single explanatory model, one can achieve a comprehensive understanding by embracing multiple complementary frameworks. This approach echoes quantum physics, where viewing phenomena through both wave and particle lenses reveals more than either perspective alone

In contemporary financial education, there is an increasing emphasis on teaching students to use a variety of analytical frameworks rather than adhering to a single methodology. This shift recognizes that quantitative and qualitative, technical and fundamental, and micro and macro approaches do not compete with each other but complement one another to build a richer understanding.

Furthermore, financial education acknowledges that these analytical frameworks are not just tools to describe financial markets. They actually actively shape and influence how those markets function. This insight calls for teaching that recognizes the dynamic, reflexive relationship between analytical models and the realities of financial systems.

Practical Application: Educational and professional institutions should develop curricula that explicitly teach complementary analytical frameworks rather than focusing on single methodologies. This means designing programs that integrate multiple disciplines, including finance, psychology, computer science, data science, and behavioral economics. This will prepare the professionals for a financial landscape characterized by increasing uncertainty, technological transformation, and system interconnection.

Future financial professionals need to combine subject expertise with various skills. Quantitative literacy will involve understanding statistics, programming, and data analysis. Behavioral insight will enable them to recognize cognitive biases and market psychology. Technological adaptability will allow them to adopt fintech innovations and digital tools. Systems thinking will build the ability to understand complex, interconnected relationships. Alongside this, they also need ethical reasoning skills for navigating stakeholder interests and social impact.

Looking back and looking ahead

As we reach the end of our quantum finance journey, it's clear that we stand at a pivotal moment in financial history. Throughout this book, we've seen how quantum principles transform every aspect of finance, from the way we understand market behavior and manage risk to how we structure corporate decisions and plan our personal financial futures.

The transformation from classical to quantum finance represents more than just new techniques or technologies. It's a fundamental shift in how we perceive uncertainty, interconnectedness, and the dynamic nature of financial systems. Where traditional finance sought to eliminate uncertainty through better forecasting, quantum finance embraces uncertainty as the foundation for creating more adaptive, resilient strategies.

For the finance professionals reading this book, the quantum revolution presents both unprecedented opportunities and urgent challenges. Institutions and individuals need to master quantum thinking. They need to learn to maintain strategic superposition, recognize measurement effects, and build adaptive systems. These enlightened professionals and institutions will thrive in the complex, interconnected markets of the future. Clinging to static, deterministic approaches expose one to forces beyond their control or prediction.

The future of finance will be written by those who understand that markets are not mechanical systems to be optimized but living ecosystems to be navigated with wisdom, flexibility, and respect for their quantum nature. Emerging tools and developing frameworks are giving early adopters a competitive advantage.

The question is not whether quantum finance will transform our industry, as that transformation is already underway. The question is whether you will be an active participant in shaping that future or a passive observer of changes beyond your control. The quantum future of finance begins with your next decision. Choose wisely, act adaptively, and prepare for a financial world that is more dynamic and interconnected and that promises greater possibilities than we have ever imagined.

Index

For Product Safety Concerns and Information please contact our
EU representative GPSR@taylorandfrancis.com Taylor & Francis
Verlag GmbH, Kaufingerstraße 24, 80331 München, Germany